The solar transition region, which spans the temperature range from 20,000 to 1,000,000 K, separates the chromosphere from the corona. All the energy that heats the corona and powers the solar wind must pass through this part of the solar atmosphere. All the mass that goes into the solar wind must flow through it.

This book summarizes recent ultraviolet and extreme ultraviolet observations of the transition region, the empirical models derived from them, and the physical models that try to explain both the observations and the empirical models. The observational focus is on quiet solar transition region observations made with *Skylab* and subsequent rocket and satellite experiments. In addition, the book presents a unified discussion of the analysis of ultraviolet and extreme ultraviolet spectroscopic data, including determination of the emission measure, and density and temperature diagnostics. This material will be useful to astrophysicists who are confronting high-resolution ultraviolet data from astrophysical plasmas for the first time.

The book will be valuable for solar physicists and astrophysicists concerned with the analysis of high-resolution ultraviolet and extreme ultraviolet observations. It could also be used as supplementary reading for graduate level courses in solar physics, stellar atmospheres or ultraviolet spectroscopy.

Cambridge astrophysics series

Editors: R. F. Carswell, D. N. C. Lin *and* J. E. Pringle

The solar transition region

Titles available in this series

1. Active Galactic Nuclei
   edited by C. Hazard and S. Mitton
2. Globular Clusters
   edited by D. A. Hanes and B. F. Madore
3. Low Light-level Detectors in Astronomy
   by M. J. Eccles, M. E. Sim and K. P. Tritton
5. The Solar Granulation
   by R. J. Bray, R. E. Loughhead and C. J. Durrant
6. Interacting Binary Stars
   edited by J. E. Pringle and R. A. Wade
7. Spectroscopy of Astrophysical Plasmas
   by A. Dalgarno and D. Layzer
10. Quasar Astronomy
    by D. W. Weedman
11. X-ray Emission from Clusters of Galaxies
    by C. L. Sarazin
12. The Symbiotic Stars
    by S. J. Kenyon
13. High Speed Astronomical Photometry
    by B. Warner
14. The Physics of Solar Flares
    by E. Tandberg-Hanssen and A. G. Emslie
15. X-ray Detectors in Astronomy
    by G. W. Fraser
16. Pulsar Astronomy
    by A. Lyne and F. Graham-Smith
17. Molecular Collisions in the Interstellar Medium
    by D. Flower
18. Plasma Loops in the Solar Corona
    R. J. Bray, L. E. Cram, C. J. Durrant and R. E. Loughhead
19. Beams and Jets in Astrophysics
    edited by P. A. Hughes
20. The Observation and Analysis of Stellar Photospheres
    by David F. Gray
21. Accretion Power in Astrophysics 2nd Edition
    by J. Frank, A. R. King and D. J. Raine
22. Gamma-ray Astronomy 2nd Edition
    by P. V. Ramana Murthy and A. W. Wolfendale
23. The Solar Transition Region
    by J. T. Mariska

The Solar Transition Region

John T. Mariska
E. O. Hulburt Center for Space Research
Naval Research Laboratory

Published by the Press Syndicate of the University of Cambridge
The Pitt Building, Trumpington Street, Cambridge CB2 1RP
40 West 20th Street, New York, NY 10011-4211, USA
10 Stamford Road, Oakleigh, Victoria 3166, Australia

© Cambridge University Press, 1992

First published 1992

Printed in Great Britain at the University Press, Cambridge

A catalogue record for this book is available from the British Library

Library of Congress cataloguing in publication data

Mariska, John T.
    The solar transition region / John T. Mariska.
      p. cm. – (Cambridge astrophysics series)
    Includes bibliographical references and index.
    ISBN 0-521-38261-0 (hc)
      1. Solar atmosphere–Spectra. 2. Sun–Corona–Spectra.
3. Emission spectroscopy. 4. Astronomical spectroscopy. I. Title.
II. Series.
    QB528.M33 1992
    523.7–dc20   92-3574 CIP

ISBN 0 521 38261 0 hardback

To Sara, Katherine, and their mother

CONTENTS

	Preface	xiii
1	**Introduction**	**1**
1.1	Historical Overview	1
1.2	The UV and EUV Solar Spectrum	4
1.3	The Outer Layers of the Solar Atmosphere	7
1.4	Scope of this Book	14
2	**Emission-Line Spectroscopy**	**17**
2.1	Atomic Processes in the Transition Region	17
2.2	Formation of Optically-Thin Emission Lines	19
2.3	Excitation	20
	2.3.1 Excitation Rates	21
	2.3.2 The Two-Level Atom Approximation	25
2.4	Ionization	26
	2.4.1 Ionization Balance	27
	2.4.2 Time Scales	31
2.5	Abundances	34
2.6	Radiation	37
2.7	Opacity	40
2.8	Diagnostics	41
	2.8.1 Temperature Diagnostics	41
	2.8.2 Density Diagnostics	43
3	**Emission-Line Intensity Observations**	**51**
3.1	Disk Observations	51
3.2	Limb Observations	56
3.3	Temporal Observations	60
3.4	Relation to Chromospheric Fine Structure	62
3.5	Relation to the Magnetic Field	67

	3.6	Relation to Coronal Structure	69
4		**Physical Conditions**	**74**
	4.1	Methods of Calculating Emission Measures	74
	4.2	Emission Measure Structure of the Transition Region	79
	4.3	Abundances	81
	4.4	Density Diagnostics	84
		4.4.1 Beryllium Isoelectronic Sequence	84
		4.4.2 Magnesium Isoelectronic Sequence	89
		4.4.3 Boron Isoelectronic Sequence	92
		4.4.4 Aluminum Isoelectronic Sequence	97
		4.4.5 Carbon Isoelectronic Sequence	98
		4.4.6 Nitrogen Isoelectronic Sequence	99
		4.4.7 Line Ratios Using Different Ions	99
		4.4.8 Transition Region Density	105
	4.5	Temperature Diagnostics	108
5		**High-Resolution Spectroscopy**	**114**
	5.1	Emission-Line Profiles	114
	5.2	Line Width Observations	116
		5.2.1 Variation with Temperature	117
		5.2.2 The C IV Line	119
		5.2.3 Relation to the Cell-Network Structure	120
		5.2.4 Center to Limb Observations	121
		5.2.5 Energetic Implications	123
	5.3	Doppler Shifts	129
		5.3.1 Observations	129
		5.3.2 Relation to Cell-Network Structure	133
		5.3.3 Relation to Line Widths	136
		5.3.4 Center to Limb Behavior	136
		5.3.5 Mass Balance	137
		5.3.6 Energy Balance	139
	5.4	Non-Gaussian Line Profiles	141
		5.4.1 Explosive Events	141
		5.4.2 Jets	149
		5.4.3 Relation to Bright Points and Spicules	150
	5.5	Temporal Variations	151
	5.6	Diagnostic Implications of a Dynamic Transition Region	157

Contents

6   Empirical Transition Region Models   **165**
- 6.1   Transforming Spectroscopic Data into Empirical Models   165
- 6.2   Effect of Inhomogeneous Structures   173
  - 6.2.1   Contribution of Spicule-Like Structures   173
  - 6.2.2   Filling Factors   176
  - 6.2.3   Absorption by Inhomogeneous Structures   180
- 6.3   Consequences of the Empirical Models   182
  - 6.3.1   Non-Maxwellian Electron Energy Distributions   182
  - 6.3.2   Diffusion   184
- 6.4   Empirical Energy Balance   185
  - 6.4.1   Global Energy Balance   186
  - 6.4.2   Local Energy Balance   187
- 6.5   Examining the Energy Balance Directly Using the Emission Measure   190
- 6.6   Observed Structure and Energy Balance   193

7   Physical Transition Region Models   **196**
- 7.1   Role of the Magnetic Field   196
- 7.2   Model Equations   200
- 7.3   Static Models   203
  - 7.3.1   Scaling Laws   203
  - 7.3.2   Static Model Solutions   205
  - 7.3.3   Emission Measures   209
- 7.4   Steady Flow Models   212
  - 7.4.1   Steady Downflow Models   212
  - 7.4.2   Loop Flow Models   213
- 7.5   Dynamic Transition Region Models   219
- 7.6   Cool Loop Models   224
  - 7.6.1   Static Cool Loop Models   224
  - 7.6.2   Combined Hot and Cool Models   228
  - 7.6.3   Radiative Losses at Low Temperatures   230
  - 7.6.4   Flows in Cool Loops   231
- 7.7   Additional Physics   232
  - 7.7.1   Diffusion   232
  - 7.7.2   Turbulent Conduction   235
  - 7.7.3   Conduction Across the Magnetic Field   236

8   The Transition Region in Perspective   **238**
- 8.1   Stellar Transition Regions   238

		8.1.1 Line Intensity Observations	238
		8.1.2 Physical Conditions and Empirical Models	241
		8.1.3 High-Resolution Spectroscopy	244
		8.1.4 Cool Stars and the Sun	246
	8.2	Physical Models and Observed Structure	248
		8.2.1 Observed Features	248
		8.2.2 Unknowns	249
		8.2.3 Model Features	250
		8.2.4 Transition Region Structure	252
		8.2.5 Future Work	253

References   256

Author Index   268

Subject Index   273

PREFACE

A few years ago, I agreed to write a review article on the solar transition region. After collecting all the recent material on the topic, I saw that there was far more than could be covered in the 25 pages I had been allotted. In addition, I realized that this portion of the Sun had been shortchanged in the past. Before the advent of space experiments, there were no observations to guide our thinking about the interface between the chromosphere and the corona. Even after rocket and satellite data began to reveal more about this part of the solar atmosphere, the topic tended to be buried in discussions of the corona or the chromosphere. This book is my effort to compensate for that past neglect.

I have chosen to emphasize transition region observations. Thus the theoretical framework may appear skimpy to some readers. There are already excellent treatments of solar magnetohydrodynamics that derive the relevant equations, explain their solutions, and present a tidier view of the structure of the outer layers of the solar atmosphere than I present here. There are, however, few up-to-date descriptions of the many high-resolution observations we now have available. Many of these observations stubbornly refuse to fit into the simple theoretical models that we try to construct to explain them.

Studying the transition region requires the analysis of ultraviolet and extreme ultraviolet spectra. I have therefore included a chapter on the formation of optically thin UV and EUV emission lines along with discussion on how they are used to diagnose the physical conditions in the transition region. While images reveal much about the nature of the transition region, spectroscopy is the key to understanding its physical properties.

The transition region also emits in the radio region in the 10 cm to 1 m wavelength range. I have chosen, however, to focus on the UV and EUV observations. With modern high-resolution radio observations of the Sun, more needs to be done to integrate the radio and UV and EUV observations into a coherent picture.

Many individuals have contributed in one way or another to this book. Numerous colleagues, both at NRL and elsewhere, willingly provided figures or data from which figures could be drawn. I am especially grateful to S. R. Habbal for helping me to navigate the photographic archives of the Center for Astrophysics and to J. F. Dowdy, Jr. for helping me obtain figures from the NASA Marshall Space Flight Center. At NRL K. P. Dere, J. W. Cook, and G. A. Doschek were always willing to help locate figures.

All my colleagues at NRL over the years have contributed to my understanding of the solar transition region, UV spectroscopy, data analysis, and numerical modelling. There are too many to mention individually, but I thank them all for making NRL such an exciting place to work. I am particularly grateful to G. A. Doschek both for his comments on the manuscript and for his encouragement over the years, and to U. Feldman, who has a fresh perspective on every topic. I take full responsibility, however, for any misstatements or errors.

While NASA has supported much of my research, the Navy has also played a key role. The management of the Naval Research Laboratory, in particular T. Coffey, has been a steady source of financial support and encouragement.

Finally, I thank my wife Pat, who allowed me to steal time from my responsibilities as husband and father to work on this project.

Burke, Virginia John T. Mariska
December, 1991

1
Introduction

In the conventional view of the solar atmosphere, the transition region separates the upper chromosphere from the corona. It has the distinction of a name because of its unusual characteristics. While the photosphere, chromosphere, and corona have gradual temperature and density gradients, the transition region appears as almost a discontinuity. In average spherically-symmetric models for the outer solar atmosphere, the temperature jumps from 25,000 to 10^6 K in only a few thousand kilometers.

Other important changes also take place in this part of the atmosphere. As the temperature rises, the atmosphere changes from predominantly neutral with radiation from hydrogen and helium dominating the spectrum to highly ionized with radiation from the less abundant heavier ions dominating. The magnetic field also changes here from being controlled by the denser photospheric gas to controlling the structure of the corona. Finally, physical processes, such as thermal conduction, which are unimportant in the photosphere and low chromosphere, become important in determining the atmospheric temperature and density structure.

1.1 Historical Overview

Even before the invention of the telescope, observers of the Sun saw the chromosphere and corona at eclipses. As the moon eclipses the last portions of the photosphere, the chromosphere becomes obvious because of its reddish appearance, the result of the strong Hα emission originating there. Visible wavelength coronal emission, which consists primarily of scattered photospheric light, has no obvious color signature. It is, however, extended and quite striking to the naked eye at an eclipse. Only recently did the question of the nature of the transition between these two distinctly different structures receive attention.

With the advent of spectroscopy and its application to the Sun and especially solar eclipses, the nature of the outer layers of the solar atmosphere became apparent. The dominance of the chromospheric emission by hydrogen emission lines made it easy to determine roughly what the temperature must be there. Emission lines were also detected in the corona at solar eclipses. Their identification, however, eluded investigators for much longer. Grotrian (1939) first suggested that the red coronal line at 6374 Å and the 7892 Å line were forbidden transitions in Fe X and Fe XI. Finally Edlén (1942) identified additional forbidden lines of highly ionized atoms and the high temperature nature of the corona became clear.

It is interesting to note that a vital clue to the identification came from stellar spectroscopy. Bowen and Edlén (1939) had discovered emission lines of highly ionized atoms in the spectrum of the nova RR Pictoris, leading Grotrian to his discovery. These astrophysical discoveries relied heavily on laboratory investigations of the spectra of highly ionized atoms. This close relationship between laboratory spectroscopy and solar spectral observations continues and has enriched both disciplines.

With the nature of the corona finally established, the details of how the temperature and density of the solar atmosphere varied from the photosphere through the chromosphere and into the corona occupied astronomers. Eclipse observations showed that large amounts of chromospheric emission extended at least 5000 km above the photospheric limb, with some emission from spicules observable to heights of 10,000 km. On the other hand, coronal emission-line observations showed that 10^6 K material extended to within at least 5000 km of the visible limb. Clearly, the details of the interface between coronal material at roughly 10^6 K and chromospheric material at roughly 10,000 K were complex.

Between the discovery of the true temperature of the corona around 1940 and the first detailed space observations in the early 1960s, theoretical speculation and models based on inadequate data provided most of the advances in understanding the nature of the transition region. Athay (1971) has provided an excellent summary of this early work. During this period, investigators developed many of the key theoretical ideas concerning the physics of the outer layers of the solar atmosphere. Giovanelli (1949), following a suggestion by Alfvén, developed a model in which a constant thermal conductive flux determined the temperature structure of the transition region. This model suggested that the temperature gradient in the transition region was steep. Unfortunately, the available observations were insufficient to allow a determination of the value of the conductive flux and hence the temperature gradient. Later, Athay and Thomas (1956) concluded, based on thermal stability arguments, that the lack of abundant efficient radiating ions formed at temperatures between He II at 40,000 K and abundant heavier ions at 10^6 K meant that temperature range should be effectively absent in the outer solar atmosphere.

During this time also theoretical work on wave and shock heating of the corona developed much of the basic physics necessary for a theoretical understanding of the transition region, including recognition of the importance of magnetic fields. Biermann (1946) and Schwarzschild (1948) first suggested that sound waves produced in the solar convection zone heated the chromosphere and corona. As they propagated upward, the waves steepened into shocks to produce the heating. Later, Osterbrock (1961) examined the role of the magnetic field in the production, propagation, and dissipation of waves. These investigations and later more detailed work provided the impetus for searches for observational evidence of the waves and stimulated many of the space experiments that followed.

Advances in our understanding of the nature of the transition region have closely paralleled advances in our ability to study the Sun in the ultraviolet with rocket and satellite experiments. Viewed as a star, the Sun has an effective surface temperature of 5770 K. Radiation from a black body at that temperature peaks at a wavelength of about 5000 Å. Thus strong continuum emission dominates the visible emission from the Sun. At 1000 Å in the ultraviolet, however, the continuum emission is reduced by a factor of roughly 10^5. This means that emission lines with intensities comparable to those seen at visible wavelengths in the corona at eclipses are observable against the solar disk in the ultraviolet. Moreover, the strong resonance lines of ions that should be abundant at transition region temperatures are located in the EUV and UV portions of the spec-

4   Introduction

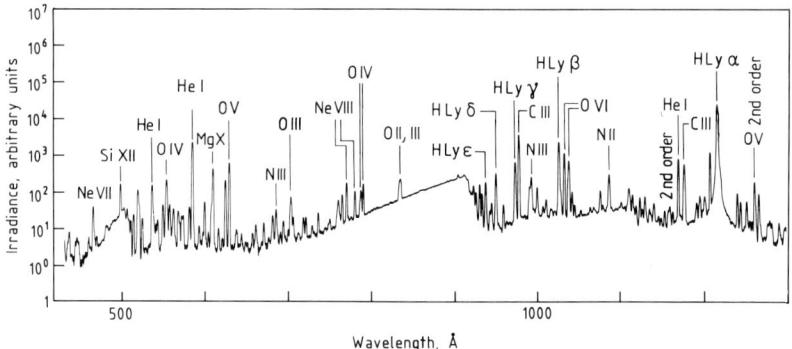

Fig. 1.1. The solar spectrum from 425 to 1300 Å recorded by the Harvard College Observatory experiment on *Skylab* and the NRL Solar Ultraviolet Spectral Irradiance Monitor on *Spacelab 2* (courtesy K. Wilhelm, Max-Planck-Institut für Aeronomie).

trum. (We define the extreme ultraviolet or EUV portion of the spectrum as extending from 100 to 1200 Å, and the ultraviolet or UV portion of the spectrum as extending from 1200 to 2000 Å.)

On October 10, 1946, using a captured V2 rocket, investigators at the Naval Research Laboratory (NRL) obtained the first UV spectrum of the Sun (Baum et al., 1946). By the early 1960s, excellent calibrated whole-disk rocket spectra were available, and their quantitative analysis began to yield important clues about the structure of the upper chromosphere, transition region, and corona. Today, highly-innovative UV experiments continue to fly on rockets. Beginning in the 1960s, UV instruments of increasing sophistication were carried into orbit, first on the *Orbiting Solar Observatory* series of satellites, then on the highly-successful *Skylab* mission, and most recently on the *Solar Maximum Mission*, and *Spacelab 2*.

1.2   The UV and EUV Solar Spectrum

Figure 1.1 shows at low spectral resolution an average quiet solar spectrum from 425 to 1300 Å. Many of the prominent emission lines are identified. Note that the intensity scale is logarithmic, so that emission lines dominate the spectrum. Only the hydrogen continuum beginning at 912 Å and the He I continuum beginning at 504 Å are comparable in intensity to the emission lines. Exami-

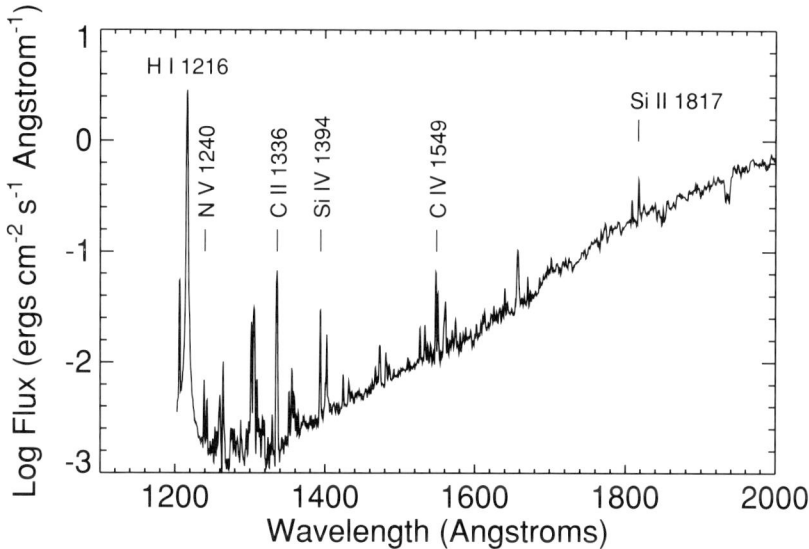

Fig. 1.2. The solar spectrum from 1200 to 2000 Å recorded by the NRL Solar Ultraviolet Spectral Irradiance Monitor on *Spacelab 2* (data courtesy M. E. Van Hoosier, NRL).

nation of the figure shows the presence of lines of all the abundant solar elements in many stages of ionization, including H, He, C, N, O, Ne, Mg, and Si.

Figure 1.2 shows at low spectral resolution how the UV solar spectrum changes at longer wavelengths. Below about 1400 Å, the spectrum continues to be dominated by strong emission lines of ions formed in the chromosphere, transition region, and corona. Above 1400 Å, however, the background continuum begins to increase relative to the emission lines, so that at Sun center only a few lines still appear in emission above 1700 Å. In the 1600–1700 Å range, the Si I free-bound continua, which are formed near the temperature minimum, dominate the continuum emission.

The spectrum shown in Figure 1.1 has a spectral resolution of about 1.6 Å, and, aside from data from a few rocket experiments, represents the highest-quality solar spectra available at wavelengths between 300 and 1150 Å. At wavelengths greater than about 1150 Å, however, spectra with a wavelength resolution of about 0.05 Å are available.

Figure 1.3 provides a better indication of how rich in detail the

6   Introduction

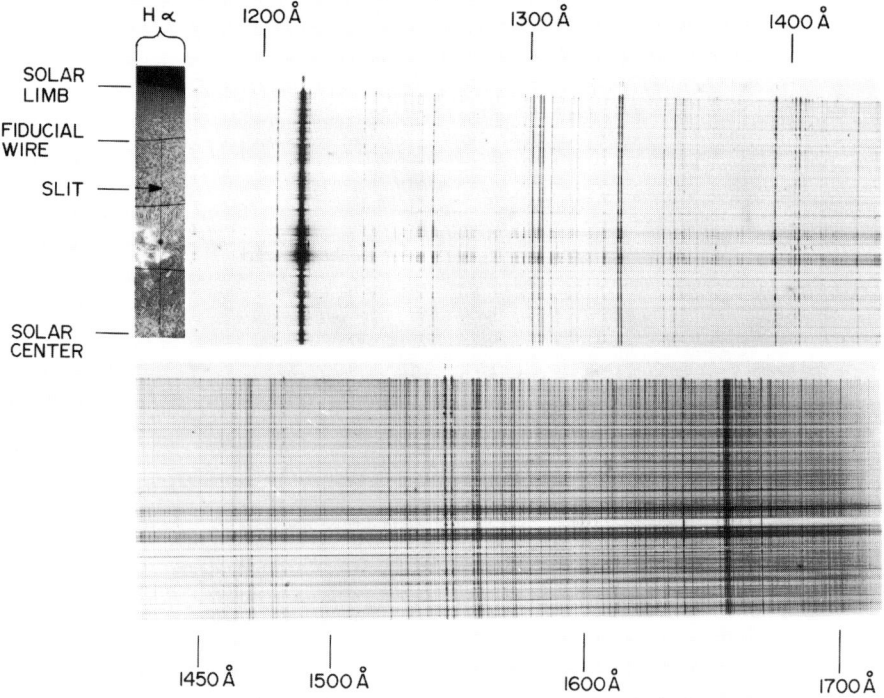

Fig. 1.3. HRTS spectrum from 1175 to 1710 Å photographed on July 21, 1975. The slit of the spectrograph lies along a solar radius from the center of the Sun to the limb, across a plage and a sunspot. An image of the Hα jaw is superposed at upper left. The vertical line through the middle of the Hα image is the 1000 arc sec long slit of the spectrograph (courtesy G. E. Brueckner, NRL).

UV spectrum is from the Lyman α line to 1700 Å. This photographic solar spectrum was obtained with the NRL High Resolution Telescope and Spectrograph (HRTS) instrument (Bartoe and Brueckner, 1975; Bartoe et al., 1977). The HRTS spectrograph is stigmatic and coma free, so there is spatial resolution along the slit. Thus the figure shows not only the character of the solar spectrum as a function of wavelength, but also the center to limb behavior.

The dominant feature in this wavelength range is the hydrogen Lyman α line at 1216 Å. At longer wavelengths the increasing strength of the background continuum is apparent. In addition, photospheric stray light is also present at wavelengths longer than 1600 Å. Thus at long wavelengths, while the spectrum is still rich in emission lines formed in the transition region, the underlying continuum increasingly contaminates the emission on the disk. Note, however, that immediately above the limb the emission lines are present uncontaminated by the underlying continuum emission. Consequently, measurements made immediately above the white-light limb permit analyses of transition region emission lines to wavelengths of about 1900 Å.

Table 1.1 lists the prominent transition region spectral lines in the EUV and UV portions of the spectrum. For wavelengths shortward of the Lyman α line at 1216 Å, the intensities listed in the table are from observations made with the Harvard College Observatory Spectroheliometer on *Skylab* (Vernazza and Reeves, 1978). These spectra were obtained with a spectral resolution of 1.6 Å and still represent the most complete set of quiet-Sun spectra from 300 Å to near the Lyman α line at 1216 Å. The spectral resolution, however, is low enough that several lines in the table are blended with other nearby lines. For wavelengths longward of the Lyman α line, the line intensities are from the NRL spectrograph on *Skylab* and the NRL HRTS experiment (Sandlin *et al.*, 1986). These data have a spectral resolution of about 0.05 Å, greatly reducing the occurrence of blends. Along with the intensities, we also list the temperature of formation in the solar atmosphere.

Examination of the temperatures listed in the table shows why this region of the spectrum is so valuable for studying the transition region and corona. Note also that, while spectral lines from many different ions are present, the same transitions keep appearing. This is because the major lines in the UV and EUV belong to only a few isoelectronic sequences. For example, the resonance lines of the Li-like ions C IV, N V, O VI, Ne VIII, and Mg X produce some of the strongest lines in the spectrum. Major transitions of the Be, B, C, N, Al, and Mg isoelectronic sequences are also present. Having many ions from the same isoelectronic sequence simplifies analyzing UV and EUV spectra because the line formation physics is similar along an isoelectronic sequence.

1.3 The Outer Layers of the Solar Atmosphere

Modern UV spectroscopic observations such as those shown in Figure 1.3 provide a hint of the complexity of the outer layers of

Table 1.1. *Prominent transition region EUV and UV emission lines. (Intensities are in erg cm^{-2} s^{-1} sr^{-1}, blended lines are marked with a b.)*

λ (Å)	Ion	Transition	$\log T_m$ (K)	Intensity
465.2	Ne VII	$2s^2\,{}^1S_0$–$2s2p\,{}^1P_1$	5.80	120
558.6	Ne VI	$2s^22p\,{}^2P_{1/2}$–$2s2p^2\,{}^2D_{3/2}$	5.65	19
562.8	Ne VI	$2s^22p\,{}^2P_{1/2}$–$2s2p^2\,{}^2D_{5/2}$	5.65	18b
609.8	Mg X	$2s\,{}^2S_{1/2}$–$2p\,{}^2P_{1/2}$	6.10	125
625.3	Mg X	$2s\,{}^2S_{1/2}$–$2p\,{}^2P_{3/2}$	6.10	51
629.7	O V	$2s^2\,{}^1S_0$–$2s2p\,{}^1P_1$	5.40	335
765.3	N IV	$2s^2\,{}^1S_0$–$2s2p\,{}^1P_1$	5.10	47b
770.4	Ne VIII	$2s\,{}^2S_{1/2}$–$2p\,{}^2P_{3/2}$	5.90	54
780.3	Ne VIII	$2s\,{}^2S_{1/2}$–$2p\,{}^2P_{1/2}$	5.90	26
787.7	O IV	$2s^22p\,{}^2P_{1/2}$–$2s2p^2\,{}^2D_{3/2}$	5.20	44
790.2	O IV	$2s^22p\,{}^2P_{3/2}$–$2s2p^2\,{}^2D_{5/2}$	5.20	83
977.0	C III	$2s^2\,{}^1S_0$–$2s2p\,{}^1P_1$	4.80	963
989.9	N III	$2s^22p\,{}^2P_{1/2}$–$2s2p^2\,{}^2D_{3/2}$	4.85	35
991.5	N III	$2s^22p\,{}^2P_{3/2}$–$2s2p^2\,{}^2D_{5/2}$	4.85	47
1031.9	O VI	$2s\,{}^2S_{1/2}$–$2p\,{}^2P_{3/2}$	5.50	305
1037.6	O VI	$2s\,{}^2S_{1/2}$–$2p\,{}^2P_{1/2}$	5.50	204b
1175.7	C III	$2s2p\,{}^3P_{0,1,2}$–$2p^2\,{}^3P_{0,1,2}$	4.80	315
1206.5	Si III	$3s^2\,{}^1S_0$–$3s3p\,{}^1P_1$	4.60	695
1218.4	O V	$2s^2\,{}^1S_0$–$2s2p\,{}^3P_1$	5.40	85
1238.8	N V	$2s\,{}^2S_{1/2}$–$2p\,{}^2P_{3/2}$	5.30	148
1242.8	N V	$2s\,{}^2S_{1/2}$–$2p\,{}^2P_{1/2}$	5.30	89
1334.5	C II	$2s^22p\,{}^2P_{1/2}$–$2s2p^2\,{}^2D_{3/2}$	4.40	309
1335.7	C II	$2s^22p\,{}^2P_{3/2}$–$2s2p^2\,{}^2D_{5/2}$	4.40	396
1393.8	Si IV	$3s\,{}^2S_{1/2}$–$3p\,{}^2P_{3/2}$	4.80	127
1402.8	Si IV	$3s\,{}^2S_{1/2}$–$3p\,{}^2P_{1/2}$	4.80	53
1548.2	C IV	$2s\,{}^2S_{1/2}$–$2p\,{}^2P_{3/2}$	5.00	207
1550.8	C IV	$2s\,{}^2S_{1/2}$–$2p\,{}^2P_{1/2}$	5.00	98

the solar atmosphere. In the next chapter we will begin to consider in detail how to use those observations to unravel the structure and dynamics of the transition region. It is useful, however, to begin with a brief overview of the conventional view of the average structure of the solar atmosphere. This will provide us with a rough indication of how the average physical conditions vary as we go outward from the visible surface of the Sun, and will provide a framework within which we can formulate some of the major questions about the structure and dynamics of the Sun's outer atmosphere. Unfortunately, beginning with an average model is also discouraging. Much of the observational material on the transition region does not integrate well into average models.

Figures 1.4 and 1.5 show at a spatial resolution of several arc sec how the appearance of the quiet Sun varies above the photosphere. At visible wavelengths the photosphere appears to be relatively homogeneous. Only the granulation gives any hint of horizontal structure. Emission from the chromosphere in for example the Ca II K line, on the other hand, shows an irregular pattern of bright emission—the chromospheric network. This network pattern is present higher in the atmosphere as seen in the He I 584 Å line, and extends through most of the transition region as seen in emission lines of He II, O V, and Ne VII.

Observations at the limb in spectral lines characteristic of the chromosphere and transition region show that this horizontal inhomogeneity extends in the vertical direction as well. Figure 1.6 shows an example of the appearance of the solar limb in the upper chromosphere and low transition region as seen in the He II 304 Å emission line, the upper transition region as seen in the Ne VII 465 Å emission line, and the corona as seen in the Mg IX 368 Å emission line. The disk portions of the images again show clearly the extension of the chromospheric network into the transition region and the more uniform coronal emission. At the limb, small-scale structuring is apparent in the He II image and to a lesser extent in the Ne VII image. The region of reduced emission at the top of the figure is a coronal hole. At chromospheric temperatures, vertical structuring is most apparent as spicules.

Magnetic field observations at photospheric levels make it clear that the distribution of magnetic flux in the outer layers correlates with the observed network structuring. These observations show

10 *Introduction*

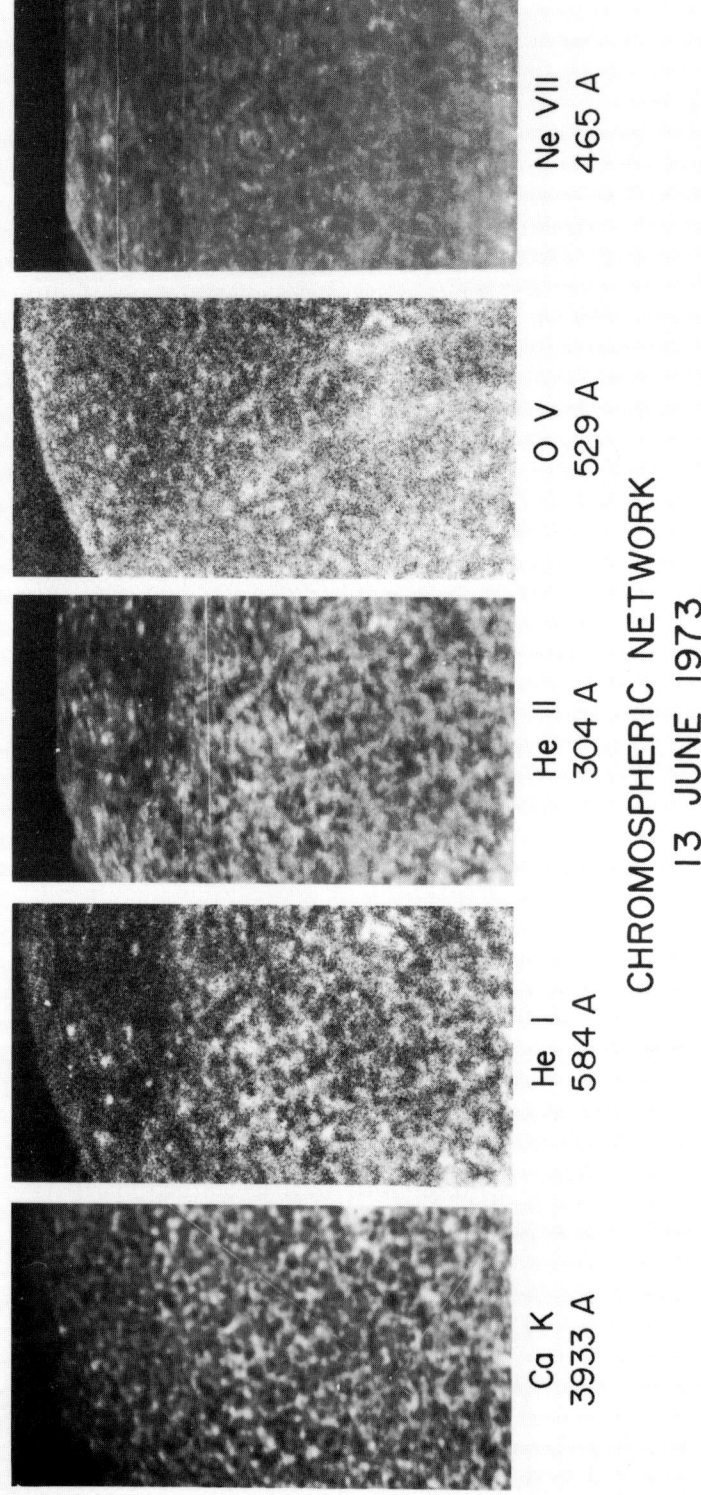

Fig. 1.4. Photographic spectroheliograms showing the appearance of the Sun from the chromosphere as seen in the Ca II K line to the upper transition region as seen in the Ne VII 465 Å emission line. The EUV images are from a calibration rocket flown during the *Skylab* mission (courtesy NRL).

The Outer Layers of the Solar Atmosphere

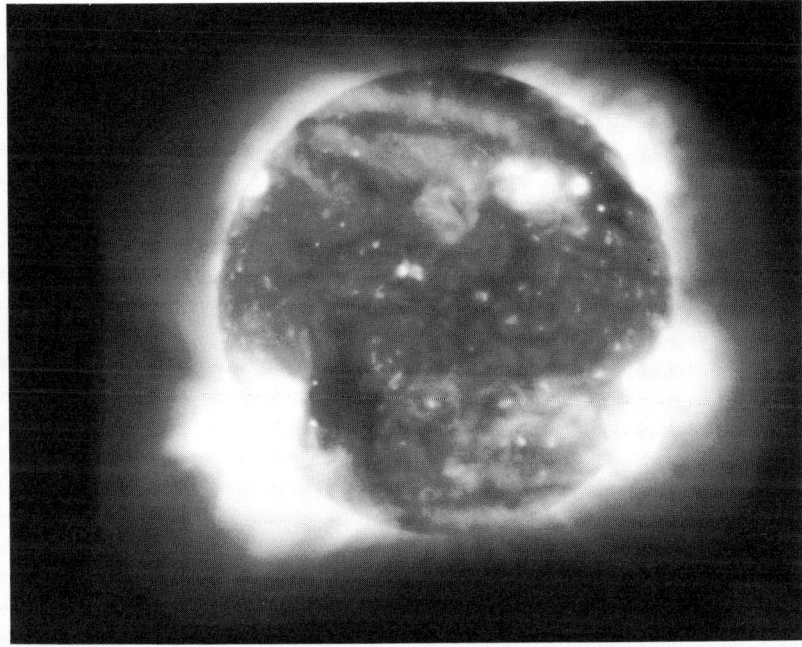

Fig. 1.5. Image of the soft X-ray corona obtained from a sounding rocket on December 11, 1987. Quiet coronal structures typical of solar minimum predominate on the disk, while active regions typical of the ascending phase of the solar cycle are present on the limb (courtesy J. D. Moses, NRL).

that the magnetic flux at photospheric levels concentrates beneath the chromospheric network.

Figure 1.7 shows an average temperature-density model for the outer layers of the quiet solar atmosphere, that is those areas of the Sun that are away from any coronal holes and active regions. The height scale begins where the optical depth at 5000 Å is unity, the usual definition of the visible surface of the Sun. From the photosphere to a temperature of 4.47×10^5 K, the model parameters are from Vernazza et al. (1981). From that temperature to a temperature of about 1.8×10^6 K, a balance between thermal conduction, radiation, and a constant volumetric energy deposition determines the model structure. Finally, above a height of about 30,000 km, the corona is isothermal. We define the photosphere to extend upward to the temperature minimum, which in

12 Introduction

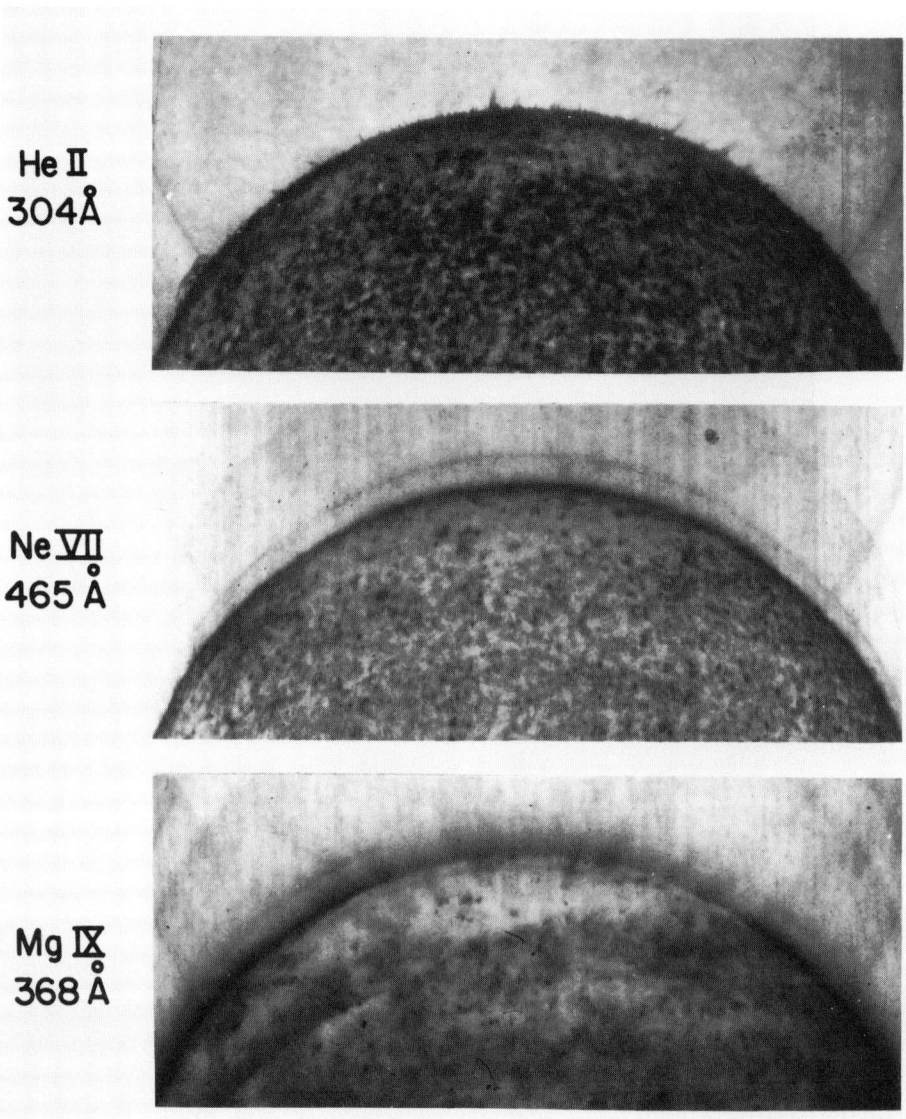

Fig. 1.6. Photographic spectroheliogram from the NRL spectroheliograph on *Skylab* showing the appearance of the solar disk and limb in the upper chromosphere, the transition region, and the corona (from Bohlin *et al.*, 1975).

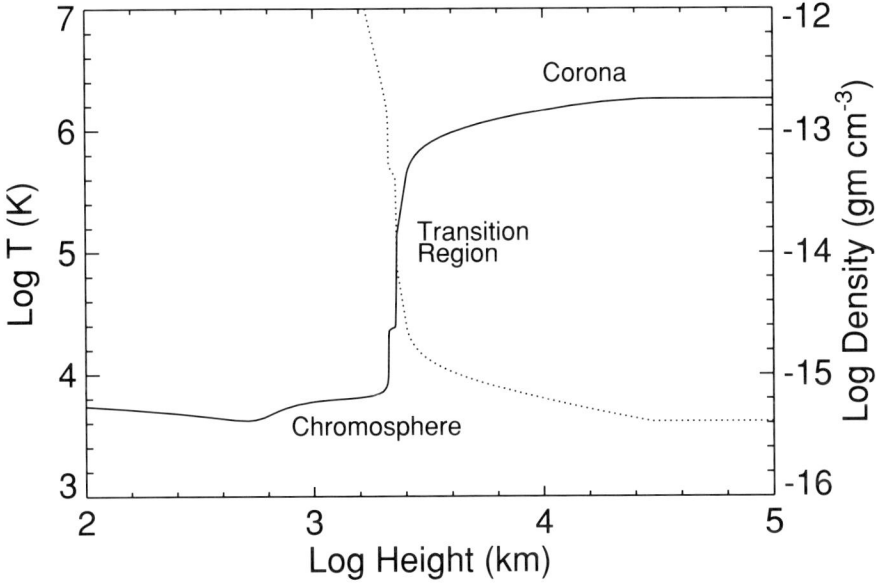

Fig. 1.7. Average temperature-density structure of the quiet solar chromosphere, transition region, and corona. The solid line is the temperature and the dotted line is the density.

this model is at a height of about 600 km. In a purely radiative-convective atmospheric model of the solar atmosphere, the temperature would continue to fall beyond the temperature minimum. It is the presence of nonradiative energy and possibly momentum deposition that causes the temperature to increase beyond the temperature minimum. The height at which this nonradiative energy deposition becomes appreciable is one way to define the boundary between the photosphere and the chromosphere. While this is an excellent physical definition, it is not very practical. We thus adopt the more common definition of the temperature minimum as the beginning of the chromosphere.

Defining the boundary between the chromosphere and the transition region is even more difficult. The most common usage, which we adopt, places this boundary at about 25,000 K, immediately above the 20,000 K temperature plateau in the model plotted in the figure. Since the temperature is already rising rapidly before reaching this boundary temperature and the actual existence of the plateau is open to question, the definition is somewhat arbitrary.

As the temperature rises above the 10^5 K height in the model, the temperature gradient begins to decrease. Eventually the temperature reaches a maximum somewhere in the corona. We will use a temperature of 10^6 K for the boundary between the transition region and the corona in quiet solar regions. It may be higher in active regions and lower in coronal holes.

Because we know the temperature at which most transition region emission lines emit, we have defined all the boundaries in terms of temperatures rather than heights. The height scale is well established in the photosphere. The white-light limb lies at a height of about 340 km above the optical depth unity point at 5000 Å (Athay, 1976), with the temperature minimum about 200 km higher. Above the temperature minimum, however, the height scale, and indeed the average model itself, become increasingly suspect.

1.4   Scope of this Book

While it has been clear since the discovery of the true temperature of the corona that maintaining the outer layers of the solar atmosphere requires nonradiative energy deposition of some kind, the nature of that energy deposition remains elusive. Acoustic waves and shocks were once the prime candidates for heating the layers above the photosphere. While they may be important in the chromosphere, magnetohydrodynamic (MHD) waves and other magnetic field related phenomena appear more likely to be important in the outer portions of the solar atmosphere. The steep temperature gradients in the average transition region model shown in Figure 1.7 suggest that thermal conduction plays an important role there. On the other hand, the average model is so idealized that it probably does not accurately represent any area on the real Sun. Thus the problem is not just one of finding the source of the heating, we must understand the flow of energy from beneath the chromosphere to the solar wind and how that energy flow produces the conditions we observe.

All the energy that heats the corona and powers the solar wind must flow through the transition region. All the mass that goes into the solar wind must flow through it. Much of the energy that reaches the corona may return through the transition region to the chromosphere by thermal conduction and as downflowing plasma.

Mass estimates for the solar wind suggest that most of the mass ejected into the corona by spicules also must return. This flow of mass and energy shapes the structure of the transition region and makes understanding this structure vital to unraveling the mass and energy balance of the outer layers of the solar atmosphere.

Our aim in this book is to summarize recent transition region observations, the empirical models derived from them, and the physical models that try to explain both the observations and the empirical models. We include as recent the data returned from the *Skylab* mission, which operated from June 1973 to February 1974. Data obtained from *Skylab* revolutionized our view of the outer layers of the solar atmosphere. Even today, more than a decade after the spacecraft itself reentered the Earth's atmosphere, analyses of *Skylab* data continue to yield discoveries. Excellent reviews of the status of transition region physics up to the time of the *Skylab* mission are already available (e.g., Athay, 1971, 1976).

While the data obtained from *Skylab* and its successors has sharpened our understanding of the physics of the transition region, there is still no unified physical model that adequately explains its observed structure. In fact, there is still considerable controversy about just what that structure is. This is true for much of the chromosphere and the corona as well. We have therefore chosen to approach the physics of the transition region from the observations. If we understand the limitations of spatial, spectral, and temporal resolution imposed by our experiments, the observations are clear. It is when we try to decode the message in the observations that the terrain becomes complex and our footing grows less certain.

Our discussion will center on the quiet Sun, though we will occasionally point out similarities and differences in the data on coronal holes and nonflaring active regions. For coronal holes, the most recent data are limited. For active regions, the available data are complex and deserve a separate detailed treatment. If, as some suggest, many features of active regions are present on a smaller spatial scale in the quiet solar transition region, then we are a long way from even a complete observational picture of the quiet transition region.

A final goal of this book is to present a unified discussion of the analysis of UV and EUV spectroscopic data. Solar physicists have pioneered the development of analysis techniques in this wavelength region. Now many other astrophysical plasmas that radiate in the UV and EUV wavelength ranges are, or soon will be, observable with spectral resolutions similar to those only solar physicists have enjoyed in the past. Our discussion of UV and EUV emission line spectroscopy aims not only at providing a basis for interpreting solar data, but also at being useful to astrophysicists who are confronting high-resolution UV and EUV data from other astrophysical plasmas for the first time.

2
Emission-Line Spectroscopy

As we saw in the average model discussed in Chapter 1, the transition region and corona are a hot, low-density plasma. If we know the temperature, density, and velocity of the plasma at each location, then we have, at least to first order, specified the thermodynamic state of the gas and can begin to investigate why it is in that particular state. This information is available to us through an understanding of the processes that excite and ionize the atoms in the plasma.

Because almost everything we know about the transition region is based on analyses of UV and EUV spectra, we must understand the line formation process in some detail. This chapter presents an overview of optically-thin emission-line formation. Its goals are to summarize the relevant atomic physics and explain the basic assumptions inherent in the analysis of UV and EUV emission lines. The material in this chapter will provide the foundation for the development of the various diagnostic techniques in Chapter 4.

2.1   Atomic Processes in the Transition Region

In a hot plasma through which radiation is propagating, many processes can excite and deexcite the energy levels of each ion and produce ionization and recombination. For UV and EUV emission lines and the conditions present in the transition region and corona, we only need to consider a few of them. Table 2.1 lists the important processes in the transition region along with the expressions for their rates and estimates of their characteristic times. The characteristic times have been calculated for the C IV ion at a

18   Emission-Line Spectroscopy

Table 2.1. *Important atomic processes in the transition region.*

Process	Rate (cm^{-3} s^{-1})	Characteristic Time (s)
Collisional excitation	$n_i n_e C_{ij}$	2×10^{-3}
Collisional deexcitation	$n_j n_e C_{ji}$	2×10^{-3}
Spontaneous radiative decay	$n_j A_{ji}$	4×10^{-9}
Collisional ionization	$n_e n_{\rm ion} q_{\rm coll}$	107
Autoionization	$n_e n_{\rm ion} q_{\rm auto}$	
Total ionization rate	$n_e n_{\rm ion} q_{\rm tot}$	107
Radiative recombination	$n_e n_{\rm ion} \alpha_{\rm rad}$	88
Dielectronic recombination	$n_e n_{\rm ion} \alpha_{\rm diel}$	
Total recombination rate	$n_e n_{\rm ion} \alpha_{\rm tot}$	88

temperature of 10^5 K and an electron density of 10^{10} cm^{-3}. For excitation processes, the resonance line at 1548 Å was used. The table does not contain characteristic times for the autoionization rate and the dielectronic recombination rate because these processes only become important in C IV at much higher temperatures.

Collisional excitations are common in the quiet solar transition region, but radiative excitations are rare due to the lack of a strong radiation field at the wavelengths of the important transitions in abundant transition region ions. Spontaneous radiative decays are important, but again because of the lack of a strong radiation field, stimulated emission is negligible. Collisional deexcitations are important in establishing level populations in some diagnostically interesting ions, but as the characteristic times in the table show, for most transitions spontaneous radiative decays will usually depopulate an excited level. For coronal forbidden lines in the visible wavelength range, the photospheric radiation field can be important.

Collisional processes are the only way to ionize most ions in the transition region, again because the radiation field is too weak to produce significant photoionization. Radiative recombination, the inverse process of photoionization, is important in recombining ions. Three-body recombination, the inverse of collisional ionization, proceeds at too slow a rate at the low densities of the transition region and corona to be important. Finally, autoionization and dielectronic recombination can play a role in determining the ionization state of the plasma. Note that the characteristic times for ionization and recombination processes tend to be in the range of tens to hundreds of seconds, compared with the much faster times for excitation processes.

2.2 Formation of Optically-Thin Emission Lines

Each atomic species contained in the plasma that comprises the transition region and corona will have many different bound energy levels. For any upper level j and lower level i, the probability of the spontaneous emission of a photon of energy $h\nu_{ji}$ per unit time is A_{ji}. If the number density of atoms in the excited level is n_j, then the volume emissivity of the plasma in the j to i transition is

$$\varepsilon_\nu = h\nu_{ji} A_{ji} n_j \psi_\nu \qquad \text{erg cm}^{-3}\ \text{s}^{-1}\ \text{Hz}^{-1}, \qquad (2.1)$$

where ψ_ν is the emission profile, normalized to unity when integrated over all frequencies. We will generally be interested in the total emissivity in the transition, which is simply

$$\varepsilon_{ji} = h\nu_{ji} A_{ji} n_j \qquad \text{erg cm}^{-3}\ \text{s}^{-1}. \qquad (2.2)$$

If we neglect opacity, the flux at the Earth from a volume of plasma ΔV is then

$$F_{ji} = \frac{1}{4\pi R^2} \int_{\Delta V} \varepsilon_{ji}\, dV, \qquad \text{erg cm}^{-2}\ \text{s}^{-1} \qquad (2.3)$$

where R is the Sun-Earth distance. In practice, the spatial resolution of the instrument we use to observe the emission line defines the volume of plasma. Since the emission is optically thin, the integral will include all the material in the line of sight from the instrument to the Sun, and, for observations above the solar limb, beyond. Thus the element of volume should be thought of as the product $A\,ds$, where A is the projected area of the instrument aperture, the solid angle, and ds is an element of the path length along the line of sight. In general, each of these volume elements may contain a complex mixture of plasma temperatures and densities. We will discuss approaches to dealing with this complication further in Chapters 4 and 6.

We can express the number density of ions in the excited level i in terms of other parameters of the solar plasma through the relationship

$$n_i = \frac{n_i}{n_{\rm ion}} \frac{n_{\rm ion}}{n_{\rm el}} \frac{n_{\rm el}}{n_{\rm H}} \frac{n_{\rm H}}{n_e} n_e. \qquad (2.4)$$

Here $n_i/n_{\rm ion}$ is the relative population of the excited level, $n_{\rm ion}/n_{\rm el}$ is the relative abundance of the ionic species, $n_{\rm el}/n_{\rm H} = A_{\rm el}$ is the abundance of the element relative to hydrogen, and $n_{\rm H}/n_e$ is the number density of hydrogen relative to the number density of electrons. Putting this relationship into equation (2.3), we have

$$F_{ji} = \frac{h\nu_{ji} A_{ji}}{4\pi R^2} \int_{\Delta V} \frac{n_j}{n_{\rm ion}} \frac{n_{\rm ion}}{n_{\rm el}} A_{\rm el} \frac{n_{\rm H}}{n_e} n_e \, dV. \qquad (2.5)$$

For an optically-thin transition region line, this equation exhibits all the parameters necessary for computing the total flux in erg cm^{-2} s^{-1} at the Earth. Our basic problem, however, is to use the observed flux in the line to derive the thermodynamic state of the plasma in which it was produced. All the factors in the integral depend, or may depend, on the physical conditions in the plasma. Thus to make use of the observed fluxes, we must understand each factor.

2.3 Excitation

As we have already shown, in a low-density plasma such as that found in the transition region and corona, the processes that populate and depopulate the excited levels of an ion are generally much faster than the processes that are responsible for ionization and recombination. This means that we can usually separate the problem of calculating the excited level populations from the problem of calculating the ionization balance. Any given level can be populated by collisional excitation from lower energy levels and by both collisional deexcitation and radiative decay from higher energy levels. The same level can be depopulated by collisional excitation to any higher level and by both collisional deexcitation and radiative decay to any lower level. A balance among these processes then determines the total population of the level.

2.3.1 Excitation Rates

Collisional transitions from level i to level j are governed by a collisional rate coefficient C_{ij} and take place at a rate $n_i n_e C_{ij}$ (cm^{-3} s^{-1}). Radiative transitions from level i to level j take place at the rate $n_i A_{ij}$ (cm^{-3} s^{-1}). So for each level in the ion, the rate equation

$$\frac{dn_i}{dt} = \sum_{j \neq i} n_j n_e C_{ji} - n_i \sum_{j \neq i} n_e C_{ij} + \sum_{j > i} n_j A_{ji} - n_i \sum_{j < i} A_{ij} \quad (2.6)$$

describes how that level's population will change with time. Here the first two terms on the right are the total rates at which level i is populated and depopulated by collisions from and to all other levels. The last two terms are the total rates at which level i is populated by radiative decays from higher levels and depopulated by radiative decays to lower levels. In practice, because of the short times for the relevant processes, we will always assume that the plasma is in a steady state and set the left side of equation (2.6) to zero. The set of rate equations for an individual ion is singly degenerate and hence can only be solved for population ratios. To calculate actual level populations, we include the requirement that the sum of the level populations results in the ion number density,

$$n_{\text{ion}} = \sum_i n_i. \quad (2.7)$$

To calculate the electron collision rate coefficient C_{ij} we integrate the cross-section for excitation by collisions with electrons of velocity v over the electron velocity distribution $f(v)$. Thus the collision rate between lower level i and upper level j is

$$n_e n_i C_{ij} = n_e n_i \int_{v_0}^{\infty} \sigma_{ij}(v) f(v) v \, dv, \quad (2.8)$$

where σ_{ij} is the electron excitation cross-section and v_0 is the velocity that corresponds to the threshold energy for the transition. For a Maxwellian distribution,

$$f(v) = 4\pi \left(\frac{m}{2\pi kT}\right)^{3/2} v^2 \exp\left(\frac{-mv^2}{2kT}\right), \quad (2.9)$$

where m is the electron mass and k is Boltzmann's constant.

It is common to see collision cross-sections expressed in terms of the collision strength $\Omega_{ij}(E)$, which is usually tabulated as a function of the kinetic energy of the exciting electron $E = \frac{1}{2}mv^2$, measured in Rydbergs. One Rydberg is the ionization energy for neutral hydrogen. The relationship between them is

$$\sigma_{ij} = \frac{\pi a_0^2 \Omega_{ij}(E)}{\omega_i E}, \qquad (2.10)$$

where a_0 is the Bohr radius and ω_i is the statistical weight of level i. Combining all this, we have

$$C_{ij} = \frac{8.63 \times 10^{-6}}{\omega_i k T^{3/2}} \int_{\Delta E_{ij}}^{\infty} \Omega_{ij}(E) \exp\left(\frac{-E}{kT}\right) dE, \qquad (2.11)$$

where ΔE_{ij} is the threshold energy for the transition. If we assume that the collision strength is independent of the incident electron energy, then

$$C_{ij} = \frac{8.63 \times 10^{-6} \Omega_{ij}}{\omega_i T^{1/2}} \exp\left(\frac{-\Delta E_{ij}}{kT}\right). \qquad (2.12)$$

If the effect of resonances is small, and we set Ω_{ij} to the value of $\Omega_{ij}(E)$ near $1.5\,\Delta E_{ij}$, then C_{ij} should be accurate to within a factor of two (Doschek, 1985).

Collision strengths must be calculated quantum mechanically using complex numerical methods. In recent years these calculations have been performed for many ions of astrophysical interest. Figure 2.1 shows an example for a typical ion of the resulting collision strength as a function of the energy of the exciting electron. Usually a collision strength has a slowly varying part upon which is superimposed a component that varies rapidly with energy. This rapidly varying component is due to the effect of resonances, which are due to dielectronic captures. Instead of the incident colliding electron simply exciting the transition and moving off with less energy, it is captured by the target ion to form a doubly excited state of the next lowest stage of ionization. This doubly excited state then immediately breaks up with the electron again released. This process only takes place within discrete kinetic energy levels of the exciting electron. Integrating the collision strength over a Maxwellian distribution averages out the effect of the exact positions of the resonances. Their contribution to the integrated collision strength can, however, be substantial. This is particularly true for intercombination and forbidden transitions, where they can enhance collision rates by a factor of two or more (e.g., Malinovsky, 1975).

Occasionally there will be transitions for which detailed collision strength calculations are unavailable. For those cases there is an

Excitation

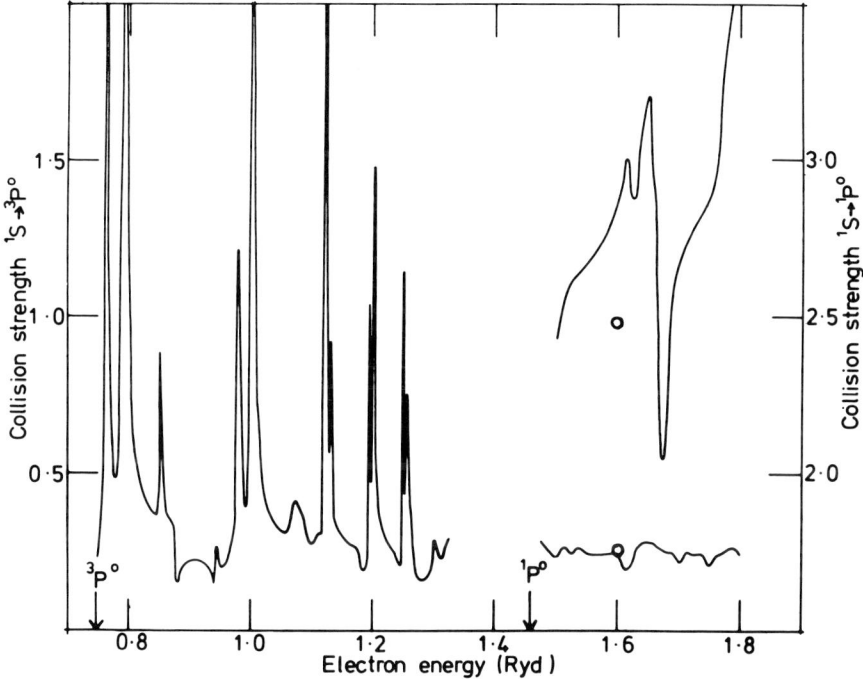

Fig. 2.1. Resonance fine structure in the collision strengths for two O v transitions (from Berrington et al., 1977).

approximate formula developed by Van Regemorter (1962). Using the Bethe approximation and comparisons with experimental and more detailed theoretical results, he obtained the approximation

$$\Omega_{ij} = \frac{8\pi}{\sqrt{3}} \frac{I_\mathrm{H}}{\Delta E_{ij}} g\omega_i f_{ij}, \tag{2.13}$$

where I_H is the ionization energy of hydrogen, f_{ij} is the atomic oscillator strength, and g is an effective Gaunt factor. For ions, Van Regemorter found g to be 0.2 for transitions in which the principal quantum number changes, while Blaha (1969) found an average value of 0.7 for transitions in which it does not. This formula is based on dipole transition selection rules and is quite approximate. Perhaps its best use is to aid in interpolating a collision strength using other known collision strengths in the same isoelectronic sequence.

Collisional deexcitation rates are obtained using the principle of detailed balance, which requires that in thermodynamic equilibrium each microscopic process be balanced by its inverse. For collisional excitations and deexcitations, this means that the number of excitations caused by electrons in a range dv_1 at velocity v_1 is balanced by collisional deexcitations by electrons in a range dv_2 at velocity v_2, where

$$\frac{1}{2}mv_1^2 = \frac{1}{2}mv_2^2 + \Delta E_{ij}. \tag{2.14}$$

Applying this principle, we have

$$n_e n_i \sigma_{ij}(v_1) f(v_1) v_1 \, dv_1 = n_e n_j \sigma_{ji}(v_2) f(v_2) v_2 \, dv_2. \tag{2.15}$$

In thermodynamic equilibrium we know from the Boltzmann equation that

$$\frac{n_j}{n_i} = \frac{\omega_j}{\omega_i} \exp\left(\frac{-\Delta E_{ij}}{kT}\right). \tag{2.16}$$

Putting all this together, we have

$$C_{ji} = \frac{\omega_i}{\omega_j} C_{ij} \exp\left(\frac{\Delta E_{ij}}{kT}\right). \tag{2.17}$$

Collisional excitation by protons also can be important. Seaton (1964) first showed that for the fine structure levels of the ground configurations of coronal ions, proton collisional excitation rates are comparable to electron collisional excitation rates. The emission lines produced by these excitations are in the visible and infrared regions of the spectrum. Examples in the solar corona include the Fe XIV green line at 5303 Å and the Fe X red line at 6374 Å. In the transition region, proton collisions affect excitation within the ground configurations of ions such as N III, O III, and O IV. While the resulting emission lines do not appear in the UV portion of the spectrum, accurate populations for all the levels are important for calculations of diagnostic line ratios.

There have been several calculations of proton collision rates. For example Faucher *et al.* (1980) have calculated rates for the ground configurations of ions in the $2p^2$ isoelectronic sequence and Bely and Faucher (1970) have published rates for the ground configurations of ions in the $2p$, $2p^5$, $3p$, and $3p^5$ isoelectronic sequences. The basic equation for the collision rate has the same form as equation (2.11) and simply results in additional terms in the excitation rate equations.

2.3.2 The Two-Level Atom Approximation

For many emission lines, such as the resonance lines of the Li-like ions, only the ground level and the excited level responsible for the line are important for calculating the line flux. In those cases equation (2.6) for the level populations reduces to the two-level atom approximation in which collisional excitations from the ground level balance spontaneous radiative decays from the excited level. Thus we can write

$$n_e n_1 C_{12} = n_2 A_{21}. \tag{2.18}$$

Frequently the populations of other excited levels in the ion are so small that all the ions are effectively in the ground level. Then the expression for the flux in the line becomes

$$F_{21} = \frac{h\nu_{21}}{4\pi R^2} \int_{\Delta V} n_e n_{\text{ion}} C_{12} \, dV. \tag{2.19}$$

This two-level atom approximation greatly simplifies calculations of emission-line fluxes because we do not have to solve the complete set of level population equations for the particular ion. Instead, we only need to know the ion number density from an ionization balance calculation.

The factors in the integral depend on both the temperature and the density. If we assume that the collision strength is independent of the incident electron energy, then we can use equations (2.4) and (2.12) to obtain

$$F_{21} = \frac{h\nu_{21}}{4\pi R^2} \frac{8.63 \times 10^{-6} \Omega_{12}}{\omega_1} 0.8 A_{\text{el}}$$
$$\times \int_{\Delta V} n_e^2 \frac{n_{\text{ion}}}{n_{\text{el}}} T^{-1/2} \exp\left(\frac{-h\nu}{kT}\right) dV, \tag{2.20}$$

where we have taken the hydrogen-to-electron number density ratio to be 0.8. Often all the temperature-dependent terms are grouped together into what is called the contribution function $G(T)$, where

$$G(T) = \frac{n_{\text{ion}}}{n_{\text{el}}} T^{-1/2} \exp\left(\frac{-h\nu}{kT}\right). \tag{2.21}$$

This function peaks sharply in temperature because of the strong temperature sensitivity of the relative ion abundances. Often we refer to the temperature at which the $G(T)$ function peaks as the temperature of formation of the line.

Before large numbers of collision strengths were available, many investigators used the Van Regemorter (1962) approximation given in equation (2.13). Incorporating this we have

$$F_{21} = \frac{2.2 \times 10^{-15}}{4\pi R^2} f A_{\text{el}} \int_{\Delta V} g G(T) n_e^2 \, dV. \tag{2.22}$$

In discussing emission lines for which the two-level atom approximation applies, we will frequently use this expression for the flux with the understanding that equation (2.20) also may be used if a measured or calculated collision strength is available.

Occasionally there will be a situation in which an excited level is populated predominantly by collisions from the ground level g, but is depopulated by radiative decays to both the ground and intermediate levels. The excitation equation for the excited level then becomes

$$n_e n_g C_{gi} = \sum_{j<i} n_i A_{ij}, \tag{2.23}$$

so that the excited level population density becomes

$$n_i = \frac{n_e n_g C_{gi}}{\sum_{j<i} A_{ij}}. \tag{2.24}$$

This then leads to the expression

$$F_{ig} = \frac{h\nu_{ig}}{4\pi R^2} \frac{8.63 \times 10^{-6} \Omega_{gi}}{\omega_g} \frac{A_{ig}}{\sum_{j<i} A_{ij}} 0.8 A_{\text{el}}$$

$$\times \int_{\Delta V} n_e^2 G(T) \, dV, \tag{2.25}$$

which reduces to the two-level case, when there are no other routes for radiative decays.

2.4 Ionization

For a low-density plasma like the transition region and corona, a balance between electron impact ionizations and radiative and dielectronic recombinations determines the number density of an ion of a particular element. Photoionization, the inverse process of radiative recombination, is unimportant in the quiet solar atmosphere, but it may be significant in active regions and particularly in flares (e.g., Nussbaumer and Storey, 1975).

2.4.1 Ionization Balance

For any given ion z of an element of nuclear charge Z, electron impact ionizations to ion $z+1$ are governed by an ionization rate coefficient q_z, and recombinations to ion $z-1$ are governed by a total recombination rate coefficient α_z. For each ion the rate equation that describes how the ion's number density will change with time is then

$$\frac{dn_z}{dt} = n_e(n_{z-1}q_{z-1} + n_{z+1}\alpha_{z+1}) - n_e n_z(q_z + \alpha_z). \quad (2.26)$$

Here the first term on the right is the rate at which ion n_z is created by electron impact excitations from ion $z-1$ and by recombinations from ion $z+1$, and the second term is the rate at which ion z is destroyed by ionizations to ion $z+1$ and recombinations to ion $z-1$. To calculate actual number densities of the ions, we require that the sum of the ion number densities results in the elemental number density,

$$n_{el} = \sum_{z=0}^{Z} n_z. \quad (2.27)$$

As we noted earlier, the time scales for ionization and recombination in the transition region and corona are in the range from tens to hundreds of seconds. These times are slow enough that departures from equilibrium are possible. In those cases equation (2.26) must be solved together with an atmospheric model, which provides the temperature, density, and velocity as a function of position and time. To make any progress, however, in understanding data from spectral lines formed in the transition region and corona, we must make simplifying assumptions about the ionization balance. The most common approximation is to assume that the plasma is in ionization equilibrium. Thus for each ion z from 0 to Z, we can write

$$n_z q_z = n_{z+1}\alpha_{z+1}. \quad (2.28)$$

Note that the electron density has dropped out of the set of equations. This means that, to the extent that we can ignore density effects in the ionization and recombination rate coefficients, the ionization balance depends only on the temperature. It can therefore be computed once and the ratio n_{ion}/n_{el} tabulated for use in interpreting observations.

Ionization can take place either by direct electron impact ionization from the ground configuration or by autoionization. In direct electron impact ionization, an incoming electron collides with an ion of charge z and produces an ion of charge $z+1$ along with another free electron. For autoionization the process is more complex. In a nonhydrogenic ion there will be bound states of the ion that lie above the ionization limit for removing an additional outer shell electron. These states correspond to the excitation of one of the inner shell electrons. If a collision from an incident electron produces one of these states, one of two things can happen. Some fraction of the time, the excited state simply decays radiatively to another bound state below the ionization threshold. Occasionally, however, the bound state interacts with the adjacent continuum state of the next highest degree of ionization, resulting in a free electron and an ion of charge $z+1$. This process is referred to as autoionization. Its net effect is the same as direct impact ionization. To include it in the ionization rate, we must multiply the cross-section for each collisional excitation to a state above the ionization threshold by the probability that the state will autoionize. Because autoionization begins with a collisional excitation, the autoionization rate varies with the same $T^{-1/2}\exp(-\Delta E/kT)$ dependence as ordinary collisional excitation rates. Collisional ionization rates, on the other hand, vary as $T^{1/2}\exp(-I/kT)$, where I is the ionization energy.

Recombination can take place either radiatively or dielectronically. In radiative recombination a free electron within the field of an ion makes a radiative transition to a bound level and a photon carries away the excess energy. Most recombinations occur to highly excited levels. Thus a hydrogenic approximation can usually be applied. In dielectronic recombination a free electron is first captured into an autoionization state of the potentially recombined ion. The system can then either autoionize back to a free electron and the ion, or can complete the recombination into a stable recombined ion by radiative transitions into bound states. In the solar corona, this process is often more important than radiative recombination.

Because they must include recombination to many different levels, the expressions for both the radiative and dielectronic rate coefficients can be quite complex. Moreover, there are only a limited number of experimental measurements. Generally one uses the

Ionization

theoretical expressions to calculate the rate coefficients and then adjusts for the experimental data.

Calculations and experimental measurements of ionization and recombination rate coefficients are scattered throughout the astrophysical, plasma physics, and atomic physics literature. Several investigators have, however, collected useful parameterizations of the rate coefficients and tabulated the resulting ionization equilibrium for astrophysically interesting ions. Ionization equilibria that include the processes of interest in the solar transition region have been published by Jordan (1969, 1970), Allen and Dupree (1969), Landini and Monsignori Fossi (1972), Summers (1972, 1974), Jacobs *et al.* (1977*a, b*, 1979), Shull and Van Steenberg (1982*a, b*), and Arnaud and Rothenflug (1985).

The compilations of Shull and Van Steenberg (1982*a, b*) and Arnaud and Rothenflug (1985) are particularly useful. They not only tabulate the ionization equilibrium, but also provide convenient parameterizations of the rate coefficients, which are useful for calculating the ionization balance using equation (2.26) in situations where the plasma is not in ionization equilibrium. Both collections are similar and yield ionization equilibria that are in good agreement. Arnaud and Rothenflug (1985) also list a number of errors in the data published by Shull and Van Steenberg (1982*a*). Recently, Landini and Monsignori Fossi (1990) also have published an excellent tabulation of parameterizations of the rate coefficients for many elements.

As an example of how ion abundances vary with temperature, Figure 2.2 shows the relative concentrations of each ionic species of oxygen as a function of temperature. The interplay between the ionization and recombination rates from ion to ion determines the shapes and maxima of the curves. At the lowest temperatures nearly all the oxygen is neutral. As the temperature increases, collisional ionization creates more O II. When the temperature becomes great enough that the ionization rate to O II is substantially larger than the recombination rate back to O I, O II becomes established as the dominant ion. The interplay between ionization to O III and recombination back to O II then becomes important. This continues from ion to ion as we move up in temperature, until eventually only fully ionized O IX is present when the temperature exceeds 10^7 K.

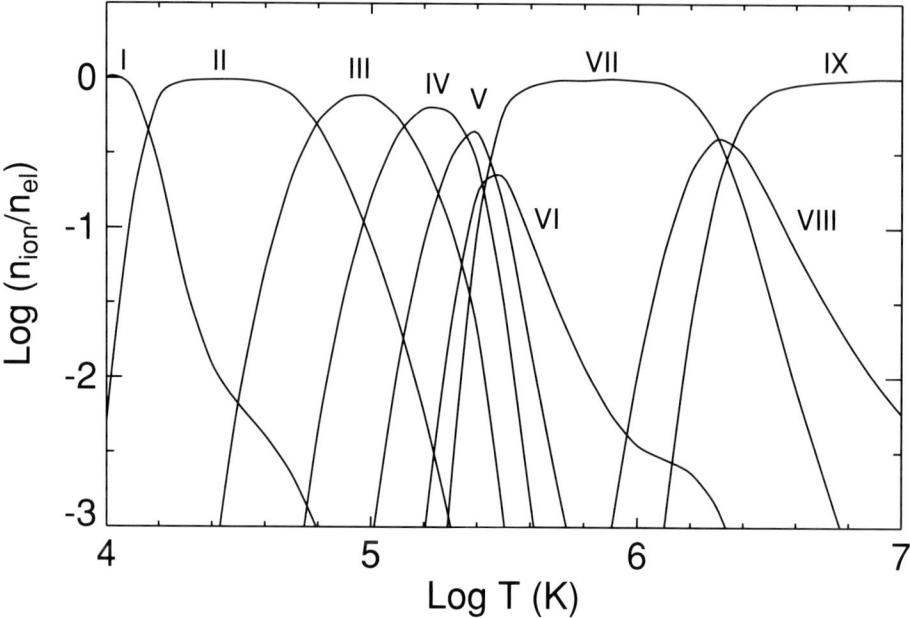

Fig. 2.2. Ionization balance for oxygen (data from Shull and Van Steenberg, 1982a, b).

The value of the maximum abundance for each ion depends to a large extent on how easy it is to ionize to the next highest stage. Ions representing closed shells, such as He-like O VII, have broad maxima with peak abundances near unity. This is because the extra energy needed to achieve the next highest stage of ionization is substantial, requiring a large temperature increase before the next ion becomes important. On the other hand, ions representing the last electron in a shell, such as the H-like O VIII and the Li-like O VI, have maximum relative abundances that are significantly below unity, because the extra energy necessary to ionize them is small.

Li-like ions, such as O VI, have not only low relative abundance peaks, but also a high-temperature tail due to dielectronic recombination from O VII. Because O VII is He-like, it exists in great abundance over an extended temperature range. In addition, again because O VII is He-like, it takes considerable energy to excite an autoionizing state of O VI from which dielectronic recombination can proceed. The net result is that the large amount of O VII leads

Ionization

to significant dielectronic recombination to O VI over a large temperature range, producing the high-temperature tail. This high-temperature tail is present on the other Li-like ions, such as C IV, and on other ions for which the next highest stage of ionization is abundant over an extended temperature range. For emission lines from Li-like ions formed in the transition region, it adds ambiguity to a determination of the temperature of the plasma emitting the line (Kozlovsky and Zirin, 1968). This is unfortunate, since the simple two-level atomic model accurately models resonance-line formation in the Li-like ions.

2.4.2 Time Scales

In an isothermal plasma of solar composition at constant density, equation (2.28) describes the equilibrium ionization balance for each element. The outer layers of the solar atmosphere, however, are highly structured and vary with time. To assess the potential for departures from ionization equilibrium, we need to evaluate typical time scales for ionization and recombination.

For an ion of charge z, the collisional ionization time to reach charge state $z + 1$ is

$$\tau_I(z) = (n_e q_z)^{-1}, \tag{2.29}$$

and the recombination time to reach the ion from charge state $z+1$ is

$$\tau_R(z+1) = (n_e \alpha_{z+1})^{-1}. \tag{2.30}$$

Table 2.2 lists values of $\tau_I(z)$ and $\tau_R(z+1)$ at several temperatures for all the ions of oxygen. The numbers were calculated using an electron density of 10^{10} cm^{-3} and the ionization and recombination coefficients of Shull and Van Steenberg (1982a). For the hotter ions, we only list ionization times if they are less than 10^{10} s.

If a volume of transition region plasma is instantaneously heated, the ionization times determine how rapidly the ionization balance reflects the change. Examination of the $\tau_I(z)$ values in the table and the temperatures of peak ion abundance in Figure 2.2 shows that generally ionization will be quite rapid if the temperature increases to much above the temperature of maximum concentration for a particular ion. For example, the temperature of maximum abundance of O IV is roughly $10^{5.2}$ K, while the temperature of maximum abundance of O V is $10^{5.4}$ K. If the temperature of a vol-

Table 2.2. *Ionization and recombination times for oxygen.*

Ion	log T (K)						
	4.8	5.0	5.2	5.4	5.6	5.8	6.0
	$\tau_I(z)$ (s)						
O I	4.6−2	1.5−2	6.9−3	4.0−3	2.7−3	2.1−3	1.8−3
O II	6.5	4.8−1	8.6−2	2.7−2	1.2−2	7.0−3	4.7−3
O III	5.8+2	1.1+1	8.6−1	1.6−1	5.0−2	2.3−2	1.3−2
O IV	8.0+4	3.3+2	9.7	9.6−1	2.1−1	7.4−2	3.6−2
O V	1.5+8	5.2+4	3.2+2	1.2+1	1.4	3.2−1	1.2−1
O VI	...	2.5+6	5.3+3	1.0+2	7.9	1.4	4.6−1
O VII	...	...	...	...	4.5+9	1.3+6	6.8+3
O VIII	...	...	...	...	...	3.9+7	8.5+4
	$\tau_R(z+1)$ (s)						
O I	2.2+1	1.6+1	1.6+1	2.1+1	3.3+1	5.5+1	9.8+1
O II	6.9	4.4	4.2	5.1	7.5	1.2+1	2.1+1
O III	4.5	2.3	1.8	2.0	2.7	4.1	6.9
O IV	2.3	1.48	1.3	1.5	2.2	3.5	6.1
O V	2.6	2.5	3.1	4.5	7.3	1.3+1	2.2+1
O VI	1.9+1	2.8+1	4.0+1	5.8+1	8.3+1	1.2+2	1.3+2
O VII	9.6	1.3+1	1.9+1	2.7+1	3.8+1	5.3+1	7.1+1
O VIII	5.7	7.9	1.1+1	1.5+1	2.2+1	3.0+1	4.2+1

ume of transition region plasma was instantaneously raised from $10^{5.2}$ to $10^{5.4}$ K, the time for the ionization to catch up would be 0.96 s. Thus for electron densities of 10^{10} cm^{-3}, ionization times to the dominant ion will generally be a few seconds at most.

Since the ionization rate depends on the electron density, ionization times can become significant in the upper transition region in coronal holes and especially in the corona and solar wind. There the flow of plasma into the low-density solar wind will eventually result in the freezing in of the coronal abundance of each ion. The density simply becomes too low for ionization or recombination to alter the ionization balance.

If the transition region has temperature gradients as steep as those suggested by the average model shown in Figure 1.7, then plasma outflows of sufficiently high velocity can result in brief departures from ionization equilibrium. This assumes that the plasma is actually flowing through the temperature gradient, changing temperature as it goes. The situation would be different if the plasma volume was in some way insulated from the surrounding temperature gradient, as might happen, for example, with spicules.

If a volume of transition region plasma is instantaneously cooled, the recombination times determine how rapidly the ionization balance reflects the change. Examination of the $\tau_R(z+1)$ values in the table shows a different pattern than was the case for $\tau_I(z)$. For many ions, such as O III and O IV, the time scales at an electron density of 10^{10} cm^{-3} are only a few seconds. These are ions for which dielectronic recombination contributes significantly to the total recombination rate. When dielectronic recombination is important, it is generally more than an order of magnitude faster than radiative recombination. Since, however, dielectronic recombination requires the formation of an excited state of the recombined ion, it requires fairly high temperatures.

If the temperature falls below the value required to create these autoionizing states, dielectronic recombination is no longer important, leaving the slower radiative recombination to operate. This is the reason, for example, that the recombination times for O II and O III show increases at the lowest temperatures compared with the values at $10^{5.2}$ K. This is also the reason the recombination times to O VI are large at high temperatures. Because O VII is He-like, it takes considerable energy to excite the lowest autoionizing state of O VI. Thus dielectronic recombination does not become important until the temperature is higher than 10^6 K. Even then, it is much less important than in some lower stages of ionization.

The relatively long recombination times for the higher stages of ionization have important implications for studies of the transition region. If we begin with a plasma at coronal temperatures and let it cool to transition region temperatures, the ionization can lag significantly. In the case of oxygen, since the longest time is for recombination to O VI, most of the plasma will remain as O VII. Once recombination to O VI takes place, recombination to the lower stages of ionization will take place relatively quickly. If the plasma is being cooled by flowing through a temperature gradient at a typical velocity of 10 km s^{-1}, a recombination time of 100 s means that the plasma volume can move 1000 km before recombination can begin to catch up with the local temperature. If there are temperature gradients in the transition region as large as those suggested by the average model presented in Chapter 1, a change of position of 1000 km would produce a significant change in the local temperature.

Of course the problem of a cooling volume of plasma is more complex than just looking at ionization and recombination time scales. If the volume element was in some way isolated from its surroundings, it would cool radiatively. It would then be necessary to calculate the time-dependent ionization balance of all the trace ions that are significant radiators to calculate the cooling rate. If the volume element is not isolated, then thermal conduction can be an important additional source or sink of energy, and the problem becomes still more complex.

2.5 Abundances

To calculate the absolute flux in an emission line using equation (2.3) we must know the abundance of the element producing the line relative to hydrogen as well as the hydrogen to electron number density ratio. These elemental abundances can be either an additional number used in the calculation or they can be the goal of the analysis of the UV emission lines. In Chapter 4 we will examine how to determine abundances from UV emission-line observations. Here we summarize current abundance data.

Abundance determinations tend to be scattered through the literature. Critical compilations of those results include those by Ross and Aller (1976), Meyer (1985a, b), and Anders and Grevesse (1989). Transition region and coronal studies frequently use the Ross and Aller (1976) compilation. They combined all the available determinations into a single set of elemental abundances for the Sun. This includes determinations from the photosphere through the corona and the solar wind and cosmic rays. Meyer (1985a), on the other hand, has produced a set of Local Galactic abundances, which agrees well with photospheric abundances for most elements, but also uses other data, such as carbonaceous chondrites,

Table 2.3. *Solar relative elemental abundances.*

		log A	
Element	Z	Photospheric[a]	Coronal[b]
H	1	12.0	12.0
He	2	10.99	11.0
C	6	8.56	8.37
N	7	8.05	7.59
O	8	8.93	8.39
Ne	10	8.09	7.55
Na	11	6.33	6.44
Mg	12	7.58	6.57
Al	13	6.47	6.44
Si	14	7.55	7.59
S	16	7.21	6.94
Ar	18	6.56	6.33
Ca	20	6.36	6.47
Fe	26	7.67	7.59
Ni	28	6.25	6.33

[a] From Anders and Grevesse (1989).
[b] From Meyer (1985b).

H II regions, and hot stars. Along with this Local Galactic set of abundances, Meyer (1985b) examined in great detail the coronal and solar wind measurements, and produced separate abundance estimates for those regions. His definition of the corona includes the transition region.

In Table 2.3 we list the abundances for the more abundant elements found on the Sun. Both the Anders and Grevesse (1989) photospheric determinations and the Meyer (1985b) coronal values are listed on a scale in which $\log A_H = 12.0$. For most of the elements listed, the photospheric values are within 0.1 dex of the solar values tabulated by Ross and Aller and the Local Galactic values determined by Meyer. The largest differences are for He, Ne, and Ar.

Of more interest are the differences between the photospheric and coronal values. Many investigations (e.g., Meyer 1985a, b; Breneman and Stone, 1985; and references cited therein) have shown that abundances measured in the corona, solar wind, and solar energetic particles differ from photospheric measurements. The differences appear to be related to the first ionization potential of the atoms. Heavy elements with first ionization potentials greater than about 9 eV tend to be underabundant in the corona by factors of 4–6. Thus atoms such as C, N, and O tend to be underabundant, while Mg, Si, and Fe are not.

The final abundance we require to compute absolute fluxes in UV and EUV emission lines is the abundance of hydrogen relative to the electrons. Unlike elemental abundances, which we usually assume are the same everywhere in the transition region and corona, this ratio is temperature dependent. Since hydrogen and helium dominate the composition of the solar atmosphere, however, they are the only elements that we need to consider. For the upper transition region and corona, we can assume that both hydrogen and helium are fully ionized. Using the coronal abundances in Table 2.3 then gives a value for n_H/n_e of 0.83. At the other extreme, Ross and Aller (1976) suggest a value for A_{He} of 10.8, giving $n_H/n_e = 0.89$. The solar abundance of helium remains highly uncertain. Thus any value for n_H/n_e from 0.8 to 0.9 is reasonable. Often 0.8 is used. At 10^5 K, most of the abundant heavy elements are doubly ionized. Even with two electrons from each ion, their abundances are low enough relative to hydrogen and helium that their contribution to this ratio is less than 1%.

Elemental abundances are also necessary for computing the mean molecular weight of the transition region plasma. This number is required to compute the mass density. The mean molecular weight is simply the mean mass per particle. Since hydrogen and helium are the major constituents, we can write

$$\mu m_p = \frac{n_H m_p + n_{He} m_{He}}{n_p + n_{He} + n_e}, \qquad (2.31)$$

where μ is the mean molecular weight in proton masses, m_{He} is the mass of a helium atom, n_{He} is the number density of He, and n_p is the number density of protons. If we assume that the hydrogen and helium are fully ionized and use the coronal abundances from the table, we find a value for μ of 0.61. The lower helium abundance mentioned above gives a value of 0.57.

2.6 Radiation

Radiation is a major energy loss mechanism in the transition region. Thus, while our primary goal in examining emission-line formation is to develop the ability to diagnose the thermodynamic properties of the plasma from measurements of small numbers of emission lines, emission-line radiation is also important for determining the energy balance of the outer layers of the solar atmosphere.

Above about 10^5 K, the solar plasma is optically thin and calculating the radiative loss rate from the plasma simply requires that we solve the ionization and excitation balance for each major element in the plasma. The radiative loss rate at any temperature is then the sum of the emitted radiation from bound-bound, bound-free, and free-free transitions at that temperature. Although the solar transition region is clearly in a dynamic state, most theoretical work on analyzing its structure and energy balance assumes for the purpose of calculating the total radiative losses that the plasma is in ionization and excitation equilibrium. Several steady state radiative loss rate calculations have been published (e.g., Cox and Tucker, 1969; Tucker and Koren, 1971; McWhirter et al., 1975; Raymond and Smith, 1977; Cook et al., 1989).

For an optically-thin equilibrium plasma the radiation rate is given by

$$E_R = n_e n_p P(T) \quad \text{erg cm}^{-3} \text{ s}^{-1}. \tag{2.32}$$

Table 2.4 lists the $P(T)$ values calculated by Raymond (1979) for a solar plasma. Rosner et al. (1978) have fitted these values with the analytic approximation

$$\begin{aligned}
P(T) &\approx 10^{-21.85} & (10^{4.3} &< T < 10^{4.6} \text{ K}) \\
&\approx 10^{-31} T^2 & (10^{4.6} &< T < 10^{4.9} \text{ K}) \\
&\approx 10^{-21.2} & (10^{4.9} &< T < 10^{5.4} \text{ K}) \\
&\approx 10^{-10.4} T^{-2} & (10^{5.4} &< T < 10^{5.75} \text{ K}) \\
&\approx 10^{-21.94} & (10^{5.75} &< T < 10^{6.3} \text{ K}) \\
&\approx 10^{-17.73} T^{-2/3} & (10^{6.3} &< T < 10^{7} \text{ K}).
\end{aligned} \tag{2.33}$$

McClymont and Canfield (1983) have suggested that below about

Table 2.4. *Radiated power function* (erg cm^3 s^{-1}) *as a function of temperature* (K) *(from Raymond, 1979).*

$\log T$	$10^{23} P_{\text{rad}}$	$\log T$	$10^{23} P_{\text{rad}}$	$\log T$	$10^{23} P_{\text{rad}}$	$\log T$	$10^{23} P_{\text{rad}}$
4.0	0.431	5.0	64.6	6.0	11.6	7.0	4.13
4.1	6.71	5.1	56.8	6.1	11.8	7.1	3.94
4.2	22.6	5.2	58.9	6.2	11.6	7.2	2.95
4.3	19.6	5.3	62.9	6.3	11.5	7.3	2.14
4.4	13.6	5.4	62.8	6.4	10.7	7.4	1.87
4.5	12.5	5.5	41.5	6.5	8.76	7.5	1.84
4.6	14.5	5.6	20.5	6.6	7.04	7.6	1.92
4.7	22.0	5.7	14.7	6.7	5.76	7.7	2.06
4.8	37.1	5.8	12.6	6.8	4.85	7.8	2.22
4.9	57.0	5.9	10.4	6.9	4.27	7.9	2.40
						8.0	2.58

10^5 K optical depth effects will reduce the radiative loss rate below the values listed in the table. They suggested that instead the radiative loss rate rises with a T^3 dependence. This would mean that below 10^5 K the radiative loss rate would be

$$P(T) = 6.46 \times 10^{-37} T^3. \qquad (2.34)$$

Athay (1986a) has suggested, however, that this correction is too large. This uncertainty about the nature of the radiative loss curve at temperatures lower than about 10^5 K is a reflection of the fact that the solar plasma in that temperature range is no longer optically thin. In the 10,000–30,000 K range, radiation from the hydrogen Lyman α line dominates, producing a local maximum at 20,000 K (e.g., Vernazza et al., 1981). At these low temperatures, the radiative loss rates depend on the details of the atmospheric structure, particularly the hydrogen ionization. Thus at low temperatures the nature of the problem of calculating radiative losses changes dramatically. Instead of the problem of an optically-thin plasma in ionization equilibrium, we are confronted with the additional complication of the full non-LTE radiative transfer problem. That problem is beyond the scope of this book.

Figure 2.3 shows the behavior of $P(T)$ as a function of temperature. While at low temperatures hydrogen dominates the radiative

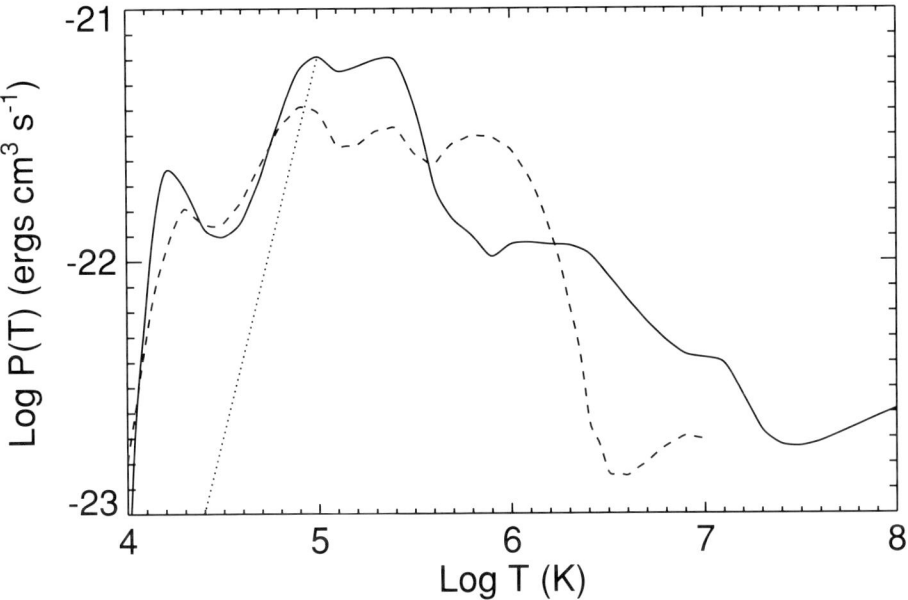

Fig. 2.3. The radiative loss function $P(T)$ derived by Raymond (1979) (solid line) based on average solar abundances and the radiative loss function derived by Cook et al. (1989) (dashed line) based on coronal elemental abundances. Also plotted as a dotted line is the T^3 radiative loss function suggested by McClymont and Canfield (1983) for temperatures below 10^5 K.

loss function, the peak in emission near 2×10^5 K results from line emission from C and O. Near 10^6 K, emissions from O, Si, S, and Fe contribute. This dominance of the radiative loss rate at transition region temperatures by a few elements means that both the absolute value of the radiative loss function and its shape depend strongly on elemental abundances. As we showed in Table 2.3, coronal abundances of C and O are significantly reduced from their photospheric values. This should lead to a change in the Raymond (1979) radiative loss function, since it is based on average solar abundances.

Cook et al. (1989) have investigated the effect of these reduced abundances on the radiative loss function. As one would expect, the reduced C and O abundances reduce the value of $P(T)$ in the 2×10^5 K region. The net result of the Cook et al. (1989) calculation was a radiative loss function with a broad flat peak that extends from about 10^5 to 10^6 K. This curve is also plotted in Figure 2.3.

At what point above the solar photosphere the depletion in the C and O abundances begins is not yet clear. There is some evidence that the abundance of O is already reduced in the transition region (Mariska, 1980). Thus the values of $P(T)$ listed in the table may not be correct for transition region calculations.

2.7 Opacity

Throughout the remainder of this book, we will usually assume that opacity will not be important in calculating emission-line intensities. Here we evaluate the optical depth in a key transition region line to show that this is usually a good assumption.

For a slab of thickness L, the line center optical depth is

$$\tau_0 = nLl_0, \tag{2.35}$$

where n is the number density of absorbers and

$$l_0 = \frac{\pi e^2}{mc} f \frac{1}{\pi^{1/2} \Delta\nu_D}. \tag{2.36}$$

Here f is the oscillator strength for the transition and $\Delta\nu_D$ is the Doppler width, given by

$$\Delta\nu_D = \frac{\nu_0}{c} \left(\frac{2kT}{M} + \xi^2 \right)^{1/2}, \tag{2.37}$$

where M is the mass of the absorbing atom and ξ is the turbulent velocity. If we consider a transition from the ground level and assume that all the ions are in that level, then from equation (2.4) we have

$$n = 0.8 A_{el} n_e \frac{n_{ion}}{n_{el}}. \tag{2.38}$$

For the C IV resonance line at 1548 Å the oscillator strength is about 0.2. Assuming an electron density of 10^{10} cm^{-3} at 10^5 K and a turbulent velocity of 20 km s^{-1}, we have

$$\tau_0 \approx 1.3 \times 10^{-8} L, \tag{2.39}$$

where we have used the coronal carbon abundance in Table 2.3 and the ionization balance calculations of Arnaud and Rothenflug (1985). Thus an optical thickness of unity will be reached when the slab has a thickness of 8×10^7 cm.

For the model shown in Figure 1.7, the thickness of the line forming region for the C IV resonance lines is about 10^6 cm. On the other hand, the highly inhomogeneous nature of the transition region may result in emitting regions that are much larger. Thus some stronger emission lines, may have some opacity, particularly near the limb, where we integrate through many structures. Often, however, we can select emission lines for analysis that have significantly smaller f-values. In addition, a small amount of opacity does not mean that photons created in the transition region will not escape. Only for values of τ_0 greater than about 10^4 will the transition region photons be trapped (Athay, 1971). Thus for all the UV and EUV emission lines we will consider, the transition region is effectively thin. This means that all the photons escape, but when there are many structures in the line of sight, some photons may be scattered out of the line of sight and hence not detected.

A small amount of opacity in an emission line will introduce additional broadening in the line profile, which can lead to erroneously large estimates of the nonthermal broadening component of the emission line. Thus we must take special care when studying nonthermal broadening to select emission lines with little or no opacity.

2.8 Diagnostics

Our primary goal in observing the UV spectrum of the Sun or any other UV-emitting astrophysical plasma is to learn the physical conditions in the observed region—to diagnose the plasma. As we have shown, both the electron temperature and the electron density enter the emission-line flux calculation. The inverse problem of determining those quantities from the fluxes is complex. Here we outline the principles involved in diagnosing the temperature and density in a UV emitting plasma. Our discussion of density diagnostics is based on an excellent treatment by Feldman and Doschek (1977b) and Doschek (1985). We will discuss actual diagnostics for the transition region in more detail in the Chapter 4.

2.8.1 Temperature Diagnostics

The temperature of an emitting plasma enters the emission-line flux equation through the temperature dependence of the relative ion abundance n_{ion}/n_{el} and through the temperature dependence of the collisional excitation rates. By far the simplest way to determine roughly the temperature of an emitting region is to look for emission lines that are primarily excited by collisions from the ground state and assume that the temperature of the maximum value of their contribution function is the plasma temperature. This estimate is quite good provided that we avoid ions that are formed over broad temperature ranges, such as Li-like, He-like, and H-like ions. It also, of course, assumes that we can use the equilibrium ionization balance, so it should be used with some caution in highly dynamic situations.

A second method for determining temperatures uses the temperature dependence in the Boltzmann factors in excitation rate coefficients. Suppose we have an ion with emission lines originating from excited energy levels 3 and 2, and ending at level 1, the lowest energy level. If we assume that levels 3 and 2 are excited from and decay to level 1, then for an isothermal plasma, the ratio of fluxes from these two lines is given by

$$\frac{F_{31}}{F_{21}} = \frac{\Delta E_{13}}{\Delta E_{12}} \frac{C_{13}}{C_{12}}. \tag{2.40}$$

Using equation (2.12) for the collision rates, we have

$$\frac{F_{31}}{F_{21}} = \frac{\Delta E_{13}}{\Delta E_{12}} \frac{\Omega_{13}}{\Omega_{12}} \exp\left(\frac{\Delta E_{12} - \Delta E_{13}}{kT}\right), \tag{2.41}$$

where we have assumed that the collision strengths are average values. If $(\Delta E_{13} - \Delta E_{12})/kT \gtrsim 1$, this flux ratio will be temperature sensitive. Of course, the solar transition region and most other plasmas are not isothermal. If the simple collisional excitation picture still applies, we then have

$$\frac{F_{31}}{F_{21}} = \frac{\Delta E_{13} \int C_{13}(n_{ion}/n_{el}) n_e^2 \, dV}{\Delta E_{12} \int C_{12}(n_{ion}/n_{el}) n_e^2 \, dV}. \tag{2.42}$$

For a plasma with regions of different temperature and density, it is possible for the two lines to be formed at different locations, resulting in a misleading diagnostic. Also, since the integrals include the relative ion abundances, changes in the ionization balance caused by flows or transient heating or cooling of a region of plasma can produce spurious results.

2.8.2 Density Diagnostics

The electron density enters the emission-line flux equation through the number density of ions in the excited level and through the density dependence of the collision rates. For a collisionally excited line from an isothermal plasma, the simplest way to estimate the density is to divide the line flux by all the temperature-dependent factors and the volume. This approach usually fails because any real plasma will have variations in both the temperature and density within the field of view of the spectrograph. Moreover, the degree of structuring is often so great that it is difficult to estimate the volume of the emitting region. Thus we obtain an average density that has little practical value, except perhaps as a lower limit.

Some ions, however, have atomic configurations that cannot be approximated as simple two-level atoms. For ions in which a two-level approximation is useful, the lines are usually produced by electric dipole transitions in which the spin does not change. These *allowed* transitions have large transition probabilities. Because of these large transition probabilities, any collisional excitation is immediately followed by a spontaneous radiative decay. But there also exist levels for which the spontaneous transition probability is smaller. Transitions from these metastable levels require a spin change, in which case they are called *intercombination* transitions, or they are magnetic dipole transitions, in which case they are called *forbidden* transitions. If an ion has one level that is always populated by collisions and depopulated by spontaneous radiative decays and a second level that is populated by collisions, but at high enough densities because of a small A-value is depopulated by both collisions and spontaneous radiative decays, then the ratio of the two emission lines produced by the spontaneous decays will be sensitive to density.

As an illustration of obtaining densities using line ratios, consider the simple three-level ion shown in Figure 2.4. Level 1 is the ground state and level 3 is metastable. The $2 \rightarrow 1$ transition is allowed, so that for all densities of interest collisions out of level 2 are insignificant. When the density is low, most of the ions are in the ground state, and collisions excite the two upper levels, which immediately decay radiatively to the ground level. Applying the

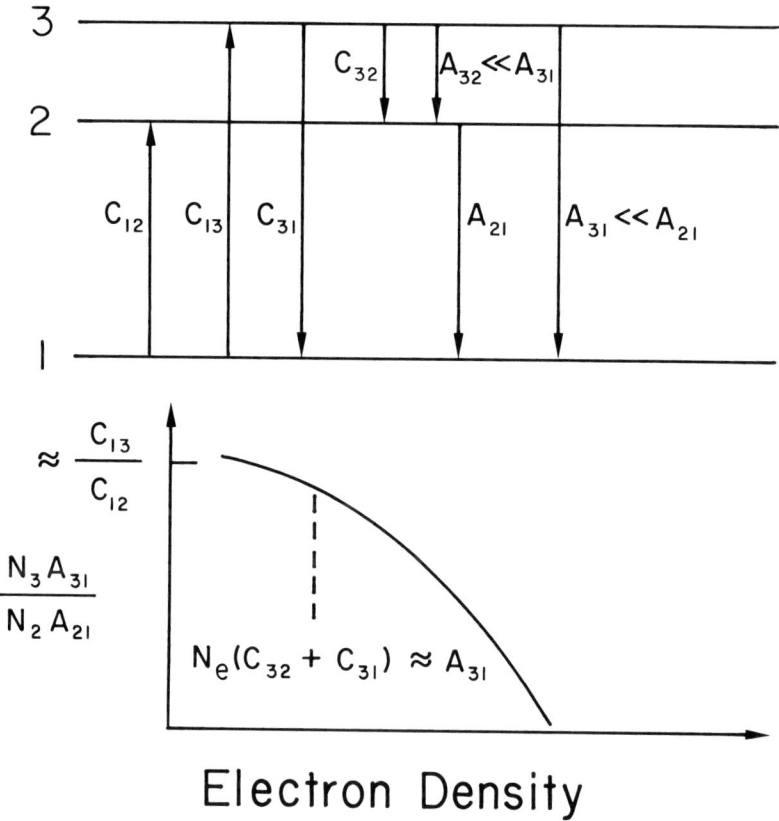

Fig. 2.4. Model three-level atom containing a metastable level, which leads to density sensitivity, and the resulting density sensitive line ratio (after Doschek, 1985).

simple two-level approximation to each transition, we find that the ratio of the line fluxes is just the ratio of the collision rates C_{13}/C_{12}.

As the density increases, however, the collisional deexcitation rates C_{32} and C_{31} become comparable to A_{32} and A_{31}. Now collisional excitations to level 3 can result in either radiative decays or collisional deexcitations. The flux in the $3 \to 1$ line then drops relative to the flux in the $2 \to 1$ line for two reasons. First, some of the collisional excitations to level 3 do not result in radiative decays, while all collisional excitations to level 2 still result in radiative decays. Second, the collisional deexcitations to level 2 result in additional $2 \to 1$ line emission. This drop in the line ratio continues until the density becomes so large that the rate of collisional excitation into and deexcitations out of level 2 exceeds the radiative decay rate. At that point the level populations are in the ratio of their statistical weights, and the line ratio is roughly proportional to A_{31}/A_{21}, which is $\ll 1$.

For the three-level case, quantifying this is straightforward. If we ignore collisional excitation from level 2 to level 3 and collisional deexcitation from level 2 to level 1, the statistical equilibrium equations for levels 2 and 3 become

$$n_3(A_{32} + A_{31} + n_e C_{32} + n_e C_{31}) = n_1 n_e C_{13} \qquad (2.43)$$

and

$$n_2 A_{21} = n_1 n_e C_{12} + n_3(A_{32} + n_e C_{32}). \qquad (2.44)$$

Combining these two equations, we have

$$\frac{n_3}{n_2} = \frac{A_{21}}{(n_e C_{32} + A_{32})(n_e C_{12} + n_e C_{13}) + n_e C_{12}(n_e C_{31} + A_{31})}. \qquad (2.45)$$

The line ratio we seek is given by

$$R = \frac{n_3 A_{31}}{n_2 A_{21}}. \qquad (2.46)$$

If we let

$$\gamma = \frac{C_{12} + C_{13}}{C_{13}}, \qquad (2.47)$$

then we have

$$R = \frac{A_{31}}{(\gamma - 1)(A_{31} + n_e C_{31}) + \gamma(A_{32} + n_e C_{32})}. \qquad (2.48)$$

When the electron density is low, the two collision terms in the denominator are small, and, since $A_{32} \ll A_{31}$, the equation simplifies to

$$R = 1/(\gamma - 1) = C_{13}/C_{12}, \qquad (2.49)$$

the expected low-density result. When the electron density is large enough that A_{31} and A_{32} are smaller than the collisional terms, we have

$$R = \frac{A_{31}}{n_e[(\gamma - 1)C_{31} + \gamma C_{32}]}. \qquad (2.50)$$

This ratio is now inversely proportional to the electron density, with only a weak temperature dependence through the temperature-dependent factors in the collision rates.

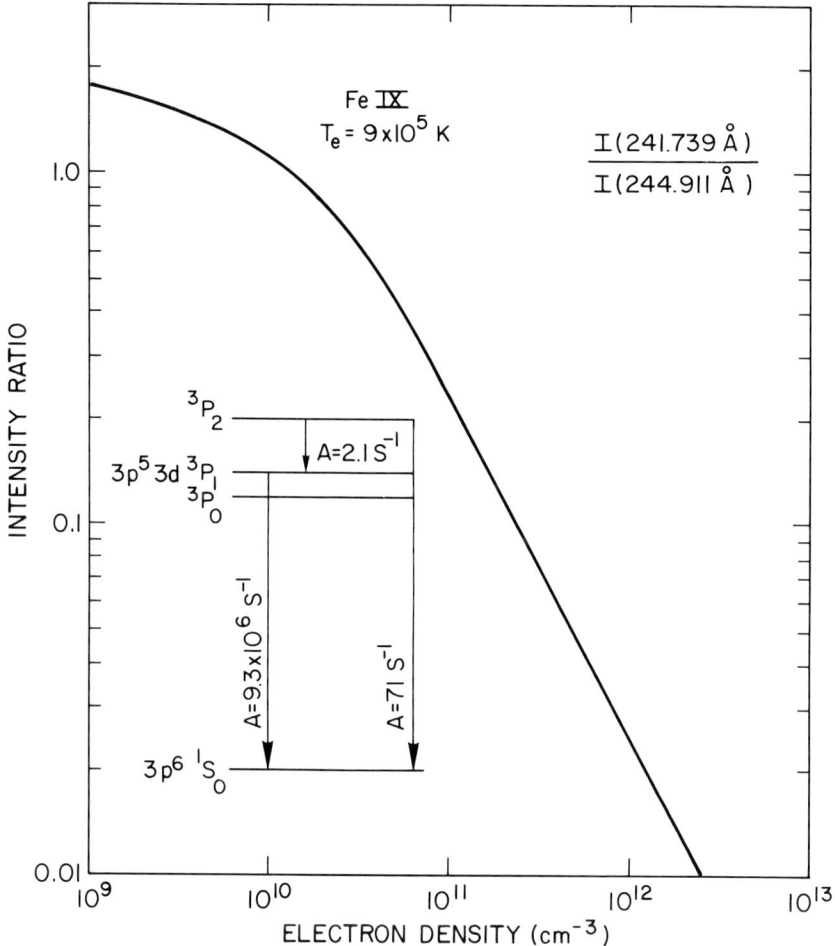

Fig. 2.5. Energy level diagram and resulting density-sensitive line ratio for Fe IX (from Feldman et al., 1978c).

Figure 2.5 shows a practical solar example of a simple three-level density diagnostic. Above the $3p^6\ ^1S_0$ ground level in Fe IX are three $3p^5 3d\ ^3P$ levels. The $^3P_1 \to\ ^1S_0$ transition at 244.9 Å is allowed and has a large A-value, while the $^3P_2 \to\ ^1S_0$ transition at 241.7 Å is a magnetic quadrupole transition, with an A-value of only 71 s^{-1}. Finally, there is a $^3P_2 \to\ ^3P_1$ magnetic dipole transition with a very small decay rate. The intensity ratio plotted as a function of density has exactly the behavior expected for the three-level case outlined.

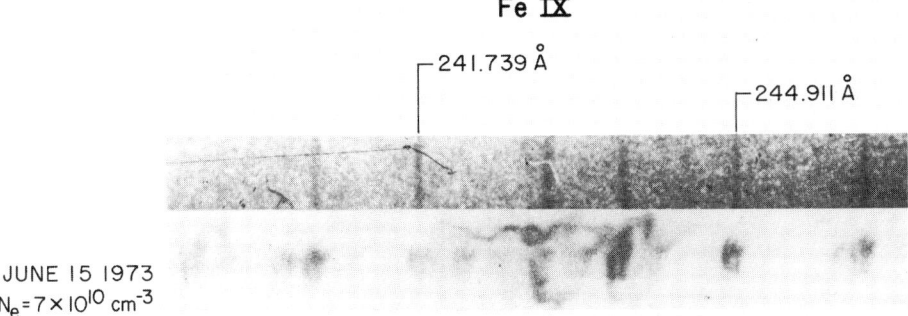

Fig. 2.6. Solar spectra illustrating the Fe IX diagnostic (from Feldman et al., 1978c).

Figure 2.6 shows an example of the application of this diagnostic. The top spectrum shows the diagnostic ratio at its low-density value, while the bottom panel shows a slitless spectroheliogram of a solar flare. At the higher densities present in the flare, the 244.9 Å line is substantially brighter than the 241.7 Å line, implying an increased density over that indicated by the top spectrum.

Figure 2.7 shows a second common configuration that yields density-sensitive line ratios. Here level 2 is the metastable level and the ratio of the flux in the $4 \to 2$ transition to the flux in the $3 \to 1$ transition is density sensitive. At low densities, only level 1 is significantly populated and levels 2, 3, and 4 are excited collisionally from it. Because of the structure of the ion, however, $C_{13} \gg C_{14}$. This would be the case, for example, if the $1 \to 3$ excitation was $^2P_{1/2} \to {}^2D_{3/2}$, while the $1 \to 4$ transition was $^2P_{1/2} \to {}^2D_{5/2}$. The collision rate for the latter transition is small because of the large change in angular momentum. Thus at low densities, the intensity ratio $R = F_{42}/F_{31}$ is proportional to C_{14}/C_{13}, which is about $1/10$. As the density increases, collisions between levels 1 and 2 begin to produce a significant population in level 2. The rate for collisional excitation from level 2 to level 4 is comparable to the rate from level 1 to level 3. This would be the case, for example, if level 2 were $^2P_{3/2}$. Since there is now an additional mechanism to populate level 4, the ratio F_{42}/F_{31} increases. This increase continues until the density becomes high enough that levels 1 and 2 reach a population determined by collisional excitation and deexcitation

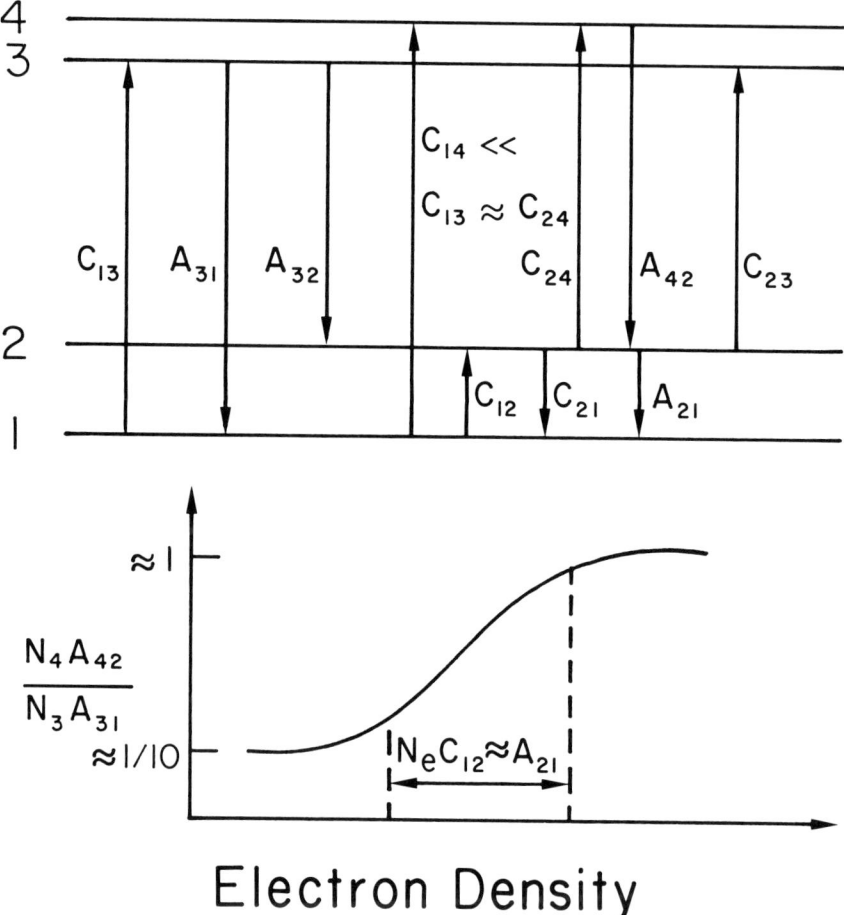

Fig. 2.7. Model four-level atom containing a metastable level, which leads to density sensitivity, and the resulting density-sensitive line ratio (after Doschek, 1985).

alone. At that point the line ratio levels off at a higher value than the low density value.

We can quantify this by considering the statistical equilibrium equations for this system. For levels 4, 3, and 1, we have

$$n_4(A_{42} + A_{41}) = n_1 n_e C_{14} + n_2 n_e C_{24}, \qquad (2.51)$$

$$n_3(A_{32} + A_{31}) = n_1 n_e C_{13} + n_2 n_e C_{23}, \qquad (2.52)$$

and

Diagnostics

$$n_1 n_e (C_{12} + C_{13} + C_{14}) = n_4 A_{41} + n_3 A_{31} + n_2 (A_{21} + n_e C_{21}). \tag{2.53}$$

Here we have ignored any mixing between levels 3 and 4. The density sensitive ratio is

$$R = \frac{n_4 A_{42}}{n_3 A_{31}}. \tag{2.54}$$

If we define

$$\alpha_{ji} = \frac{A_{ji}}{A_{j1} + A_{j2}}, \tag{2.55}$$

then from the first two statistical equilibrium equations we have

$$R = \frac{\alpha_{42}[C_{14} + (n_2/n_1)C_{24}]}{\alpha_{31}[C_{13} + (n_2/n_1)C_{23}]}. \tag{2.56}$$

The only density dependence is in the ratio n_2/n_1, which is given by

$$\frac{n_2}{n_1} = \frac{C_{12} + C_{13}\alpha_{32} + C_{14}\alpha_{42}}{A_{21}/n_e + C_{21} + C_{23}\alpha_{31} + C_{24}\alpha_{41}}. \tag{2.57}$$

For low densities n_2/n_1 is small and we have

$$R = \frac{\alpha_{42} C_{14}}{\alpha_{31} C_{13}}. \tag{2.58}$$

For large densities, the density-dependent term in the expression for n_2/n_1 becomes small and the ratio is again constant. Between these two extremes, the ratio is density sensitive.

These two cases cover most of the density-sensitive line ratios encountered in the outer layers of the solar atmosphere. In both of them, there is, of course, also some sensitivity to temperature. For the three-level case, this sensitivity is minimized if the separation in energy between levels 2 and 3 is small compared with their separation from level 1. For the four-level case, levels 3 and 4 should be close in energy and well separated from levels 1 and 2, which also should be close in energy. Even under these conditions there will always be some temperature dependence from Boltzmann factors with different energies in them.

For real density-sensitive systems, it is often necessary to solve the full set of statistical equilibrium equations instead of the reduced sets presented here. Ultimately the accuracy of a density determined using a line ratio depends on how accurately we know the excitation rate coefficients and radiative transition probabilities for all the relevant transitions, especially the weak intercombination and forbidden transitions. Continued improvements in laboratory measurements and calculations of these quantities should result in better and more consistent density determinations.

In Chapter 4 we will look at density diagnostics that are applicable to the transition region. We also will examine other methods for determining the density that depend on emission lines from more than one ion.

This completes our overview of optically-thin line formation. We have in equation (2.5) a deceptively simple looking expression for the flux in an optically-thin emission line. In principle, observations of the spatial distribution of emission in selected UV and EUV emission lines will yield a clear picture of the temperature and density distribution of the plasma in the transition region. In practice, the problem is more complex. In the next chapter we begin solving it by examining what we can infer directly from transition region observations.

3
Emission-Line Intensity Observations

Observations of the total intensity in UV and EUV emission lines constitute the largest single kind of data available for the study of the transition region. Since each emission originates in a narrow temperature range, observations in different spectral lines reveal the structure of this portion of the atmosphere at different temperatures. This chapter summarizes what we know about the morphology of the transition region at both low and high spatial resolution.

3.1 Disk Observations

When observed at even moderate spatial resolution in UV and EUV emission lines, the quiet transition region and corona show evidence for structuring. Figure 3.1 shows a typical example of a quiet portion of the surface in EUV emission lines formed in the transition region and corona. Each spectroheliogram is 5 arc min on a side (2.18×10^5 km), with the individual horizontal scan lines separated by 5 arc sec (3600 km). For all but the Mg X spectroheliogram, the same network pattern that is observed in the chromosphere dominates the image. This pattern persists, except for changes in contrast, in all transition region lines from Lyman α, which is formed at roughly 20,000 K, through the O VI 1032 Å line, which is formed at about 3×10^5 K. Between the O VI line and the Mg X line, which is formed in the corona at about 1.4×10^6 K, the regular pattern disappears. Some bright network features are traceable from Lyman α through the Mg X line, but the reverse is not always true. Observations by Brueckner and Bartoe (1974) show

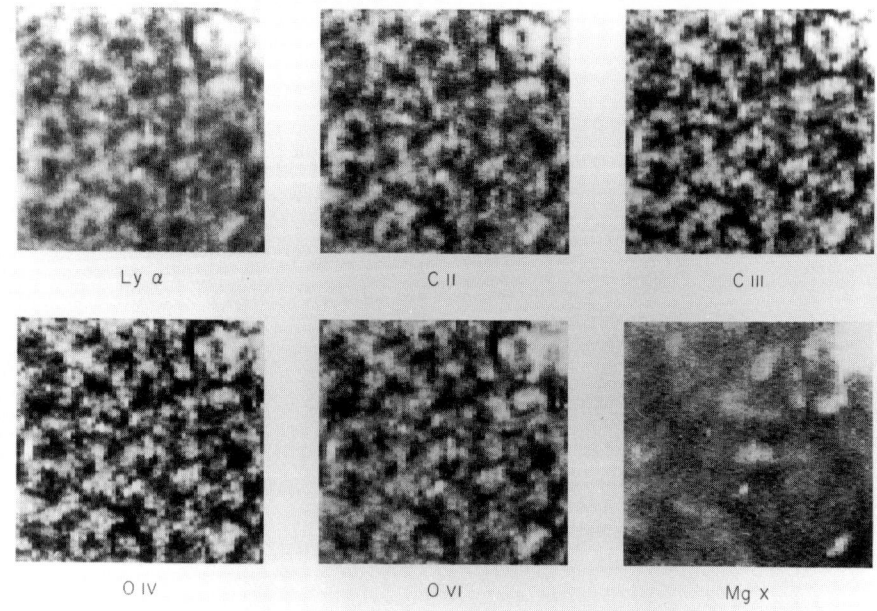

Fig. 3.1. The quiet transition region network observed approximately midway to the limb in the 1216 Å Lyman α line, the 1335 Å C II line, the 977 Å C III line, the 554 Å O IV line, the 1032 Å O VI line, and the 625 Å Mg X line by the Harvard experiment on *Skylab* (from Reeves, 1976).

that the regular network pattern is noticeably less concentrated in the Ne VII 465 Å line, which is formed at roughly 5×10^5 K.

Comparisons of EUV network observations with simultaneous or nearly simultaneous observations of the chromospheric network in the Ca II K line and in Hα show that they are cospatial (Brueckner and Bartoe, 1974; Reeves *et al.*, 1974). The EUV emission also concentrates at the location of the dark mottles typically observed in broad-band or off-band Hα filtergrams. There appears, however, to be no strong correlation between the visibility of the Hα mottles and the brightness of the EUV network (Reeves *et al.*, 1976).

While the general network appearance is similar in all the UV and EUV lines formed at temperatures below about 6×10^5 K, the

Disk Observations

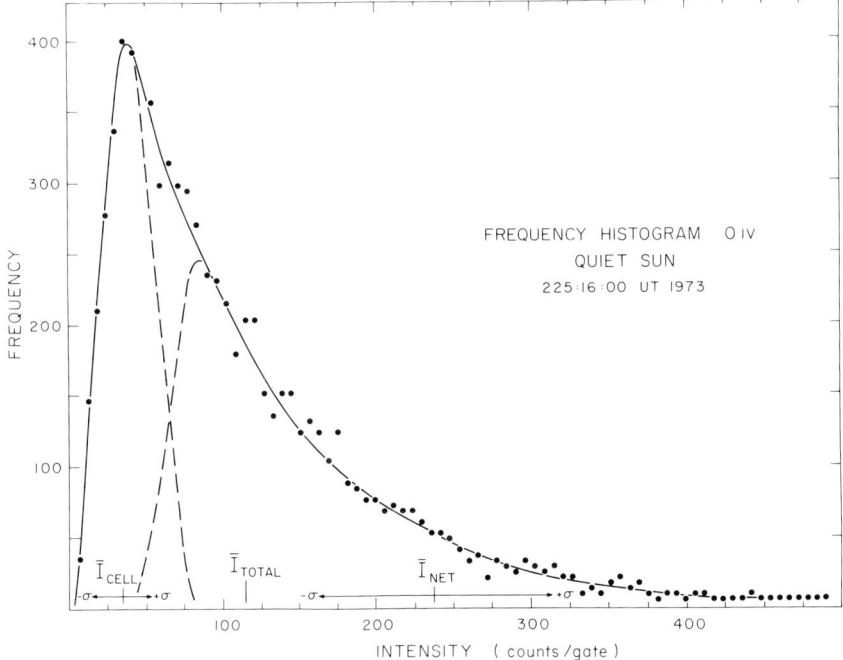

Fig. 3.2. The distribution of intensities in the O IV 554 Å emission-line spectroheliogram shown in Figure 3.1. Dashed lines show the cell-center and network contributions (from Reeves, 1976).

figure shows that both the contrast between the bright and dark regions and the general intensity distribution are changing from line to line. Reeves (1976) and Schrijver et al. (1985) have performed the most detailed studies of these and other characteristics of the EUV network and intervening cells. Figure 3.2 shows the distribution of intensities in the O IV 554 Å emission-line spectroheliogram shown in Figure 3.1. Although the spectroheliograms appear to display a clear separation into bright network elements and dark cell interiors, the intensity distribution shows no indication of a bimodal distribution. Instead, there is a peak at low intensities and an extended high-intensity tail.

Frequency distributions for other EUV emission lines show the same general appearance through at least the temperature of formation of the O VI 1032 Å emission line (3×10^5 K). In the Mg X 625 Å line, which is formed at coronal temperatures, the peak at low intensity remains, but the extended tail is much less apparent. Reeves (1976) found that if he used the average intensity in a spectroheliogram, shown as $\bar{I}_{TOTAL}$ in the figure, to construct isophote contours in the spectroheliogram, then cell interior and network areas naturally emerged. Large contiguous areas with intensities below the average correspond to the cell interior regions, while areas of higher than average intensity group together to form a network that corresponds with that seen in the Ca II K line.

When only those portions of the spectroheliogram with intensities below the average intensity are considered, the average intensity, labeled $\bar{I}_{CELL}$ in the figure, is close to the maximum in the frequency distribution, suggesting that the cell interior intensity distribution may be symmetric about the peak, with the additional skewed component due to the network. This interpretation of the chromospheric network intensity distribution observed in the Ca II K line was first suggested by Jensen and Orrall (1963) and further refined by Skumanich et al. (1975). Reeves (1976) examined the statistical distribution of the intensities and concluded that they nearly followed a normal function. The standard deviation, however, exceeds the value expected from counting statistics, suggesting structuring in the cell interiors at spatial scales less than the 5 arc sec resolution of the spectroheliograms.

As the extended tail on the intensity distribution in Figure 3.2 suggests, the network is less easy to describe using average properties. Reeves (1976) found significant intensity fluctuations along isolated network elements at the 5 arc sec resolution of his data. Moreover, one emission line may show an extended structure while another line formed at a different temperature may show smaller bright patches separated by dark gaps at the same location. One property that has been roughly quantified is the average width of a network element. From an examination of 116 network elements, Reeves (1976) found an average full width at half-maximum intensity in the 554 Å O IV of 10 arc sec, with a tail extending to 30 arc sec. Within the errors of the measurement, this average width was constant in emission lines formed at temperatures from 10,000 to 30,000 K. There was some evidence for greater average widths in the Ne VIII 780 Å emission line.

At a spatial resolution of roughly 1 arc sec, Dere et al. (1987) found somewhat different sizes for bright emitting elements in the

C IV 1548 and 1550 Å lines. They found that discrete elements along the slit of their spectrograph had an average full width at half-maximum intensity of 2450 km, about 3.4 arc sec. At lower temperatures, the width seen in the C I intercombination lines at 1630 Å was 2000 km. This difference in size estimates is probably due to selection effects. Reeves (1976) selected network elements to examine further. Dere et al. (1987), on the other hand, simply studied the discrete structures along the slit of their spectrograph. Their value is therefore probably more indicative of the size of the individual structures at transition region temperatures.

While the size of network elements at 5 arc sec resolution is roughly constant throughout the transition region, the contrast between network emission and cell emission varies considerably. There are, of course, sizable variations in measurements of the contrast because of the wide range of intensities in the network itself. Generally, however, the measurements show that the contrast between representative network elements and adjacent cell interiors has a value of between 1.5 and 2 for emission lines formed near 20,000 K, rises to a value between 2 and 6 for emission lines formed at temperatures near 2×10^5 K, and falls to near unity for lines formed at temperatures above 10^6 K (Reeves, 1976; Feldman et al., 1976b). Observations at high spatial resolution in C I and C IV emission lines show similar trends (Dere et al., 1987).

Since the contrast between network elements and cell interiors changes with the temperature of formation of the emission lines, the relative intensity contribution of the network to the average emission in each line also must change with temperature. Using spectroheliograms obtained in the O IV 554 Å line to define network elements, Reeves (1976) found that the network contributes about 60% of the quiet Sun intensity in lines formed at temperatures near 10,000 K, 75% for lines formed near 2×10^5 K, where the contrast is largest, and about 50% for lines formed above 10^6 K. Observations at 1 arc sec spatial resolution in the C IV resonance lines seem to bear out these percentages. Dere et al. (1984) found that 16% of the area along their spectrograph slit accounted for 50% of the total intensity and that 50% of the points provided 85% of the total intensity.

Estimates of the fractional area of the quiet Sun occupied by network depend on how network is defined. Using the average intensity in the entire spectroheliogram as the lowest network intensity gives a value of 50% for the area occupied by the network in O IV 554 Å spectroheliograms (Reeves, 1976). Schrijver et al. (1985) used a more complicated measure of the area. For each spectroheliogram, they first determined the symmetrical intensity distribution attributable to cell interiors. The ratio of the area responsible for those intensities to the total area then gives the fractional area occupied by cell interiors. With this method they found that the network occupied roughly half the area in the Lyman α line and in emission lines of C II and O IV, and 70% of the area in emission lines of C III and O VI. The C III line results appear to be anomalous, since its temperature of formation is between that of the C II line and the O IV line.

3.2 Limb Observations

While observations near Sun center in UV and EUV emission lines provide considerable information about the surface distribution of emitting structures throughout the transition region, they provide little information about the vertical extent of the emission at each temperature. Near the limb we are able to see the vertical extent of the emission from each spectral line. Figure 3.3 shows the appearance of a quiet solar limb at 5 arc sec resolution in several EUV emission lines formed in the upper chromosphere, the transition region, and the corona. Near the limb, the bright network emission becomes increasingly important in the total emission, due to projection effects and the fact that the emitting lines are optically thin or nearly so. At the limb, the intensity in the optically-thin lines doubles because the path length doubles once the line of sight is above the opaque lower atmospheric layers. Even at 5 arc sec resolution, the chromospheric and transition region emission lines show a ragged limb, suggesting inhomogeneous structure. Some of these emitting structures extend to considerable heights above the bulk of the emission at the limb.

The small number of UV observations at a spatial resolution near 1 arc sec show a wealth of discrete emitting structures at the limb. Figure 3.4 shows an example of a high spatial resolution C IV filtergram obtained during the *Spacelab 2* mission. This filtergram

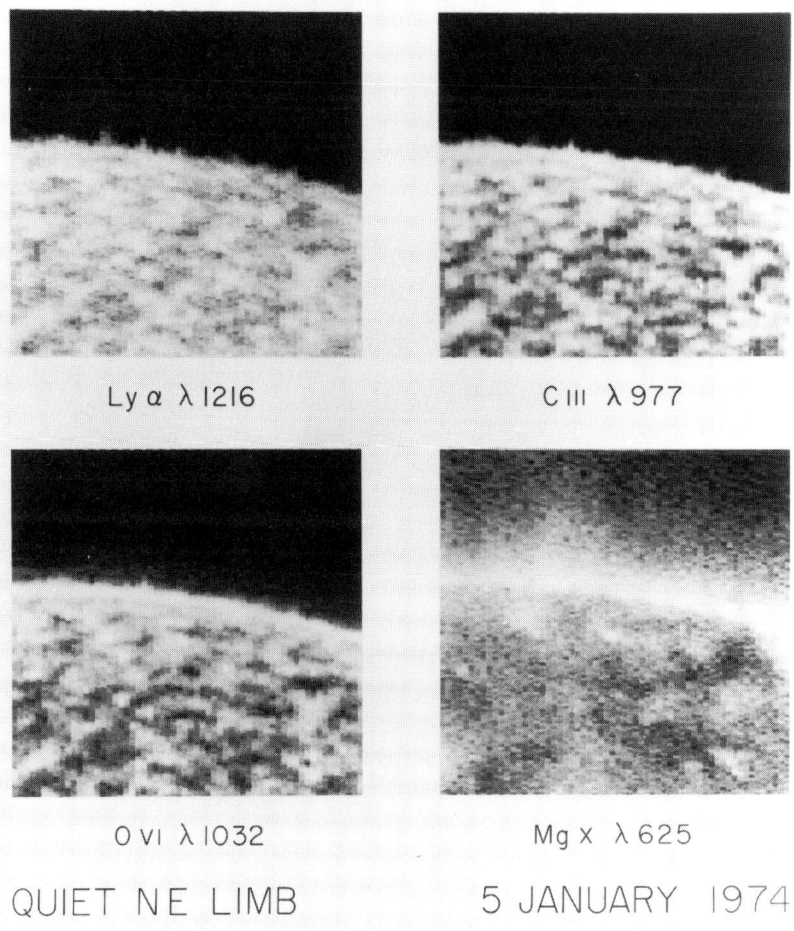

Fig. 3.3. The quiet transition region and corona at the limb observed with the Harvard spectroheliometer on *Skylab* (from Mariska and Withbroe, 1975).

covers a 90 Å region of the solar spectrum centered at 1550 Å, the location of the two strong C IV resonance lines. On the disk the bandpass contains emission from temperature minimum continua, chromospheric emission lines, and the strong C IV lines. Above the limb, however, emission from the C IV lines dominates. Well above the limb, the emission separates into narrow jets, which look very much like the spicules seen in chromospheric lines. Immediately below the UV image is an Hα image obtained simultaneously.

Even the low-resolution observations can provide some additional

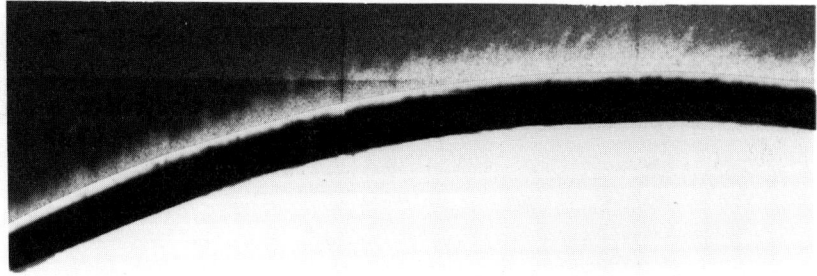

Fig. 3.4. HRTS spectroheliogram showing UV structure at the limb. The upper positive image shows C IV resonance line emission above the limb. The horizontal line in the positive image marks the boundary between the upper slit jaw, which was highly reflective, and the lower slit jaw, which was less reflective to suppress emission from the disk. Below the positive image is a simultaneous negative Hα image also obtained with the HRTS (courtesy J. W. Cook, NRL).

information on the nature of the structuring at the limb. For an optically-thin emission line, the intensity should double at the limb at the location where the line of sight goes above the absorbing continua associated with the temperature minimum region. At greater heights above the limb, the intensity will continue to rise until it peaks at the line of sight that intercepts the largest amount of emitting material for the particular line. If the transition region was spherically symmetric, the height of peak emission would correspond to the lower boundary of the spherical shell in which the particular emission line was formed. By measuring the height above the limb of the emission peak for optically-thin lines formed at temperatures throughout the transition region, it should be possible to develop a picture of the distribution of transition region material with height. Doschek et al. (1976b), Kjeldseth Moe and Nicolas (1977), and Mariska et al. (1978) have done this using data from the Naval Research Laboratory spectrograph on *Skylab*. This instrument had a 2 arc sec × 60 arc sec slit, which could be oriented tangent to the solar limb, providing 2 arc sec height resolution.

Figure 3.5 shows a typical set of limb brightening curves for optically-thin emission lines obtained with the NRL spectrograph on *Skylab*. All the quiet solar limb observations show the same characteristics for optically thin lines. Above the limb, the height at which the emission peaks increases with increasing temperature

Limb Observations

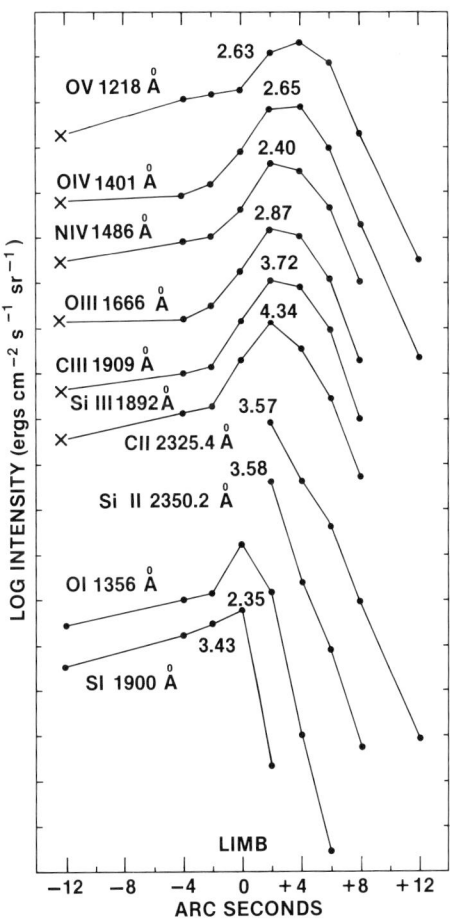

Fig. 3.5. Mean limb-brightening data for two sets of limb spectra obtained with the NRL spectrograph on *Skylab*. The number next to each data set is the logarithm of the total line intensity at +2 arc sec except for the S I 1900 Å line, where it is the logarithm of the intensity at 0 arc sec. Each division on the ordinate is 0.3 dex (from Mariska *et al.*, 1978).

of line formation. Emission from optically-thin lines of S I, O I, Si II, and C II peaks at the position of the white-light limb, while the emission from lines of higher temperature ions peaks between 2 and 4 arc sec above the limb. Since the spectrograph slit was 2 arc sec wide, the actual emission peak in these transition region lines could range from 1 to 5 arc sec above the limb. In models for the photosphere and chromosphere, the white-light limb is at a height of 340 km above the location of the optical depth unity point at 5000 Å (Athay, 1976), the zero height in the Vernazza et al. (1981) atmospheric models. Thus the bulk of the emission in lines formed between 40,000 and 2.2×10^5 K is located between 725 and 3625 km above the white-light limb, or between 1065 and 3965 km above the optical depth unity point in the atmosphere.

At heights above the peak in emission, the intensity falls rapidly. Mariska et al. (1978) measured scale heights of 1450 km for the decrease in the emission from the O V 1218 Å line and 1360 km for the O IV 1401 Å line for one quiet limb. For lines formed in roughly the same temperature range, Withbroe (1983) found an average scale height of 2000 ± 300 km. His data were obtained with the Harvard spectroheliometer on *Skylab*, which had 5 arc sec spatial resolution. The measurements also extended to greater heights above the limb than the NRL observations, suggesting that the scale height for the emission decline in transition region lines may increase at large heights above the limb.

Both the high- and low-resolution observations at the limb clearly show that a simple laminar average model such as the one presented in Figure 1.7 is inadequate. The high-resolution images show small-scale structures extending well above the bulk of the transition region emission, strongly suggesting that a significant fraction of that emission also comes from small jet-like structures. The NRL slit spectrograph observations show that the bulk of the transition region emission between 40,000 and 2.2×10^5 K spreads over at least 1500–3000 km, rather than the roughly 100 km suggested by average models.

3.3 Temporal Observations

Emission from lines formed in the transition region varies not only in space, but also in time. Changes in the emission with time at a given location on the Sun must be due to changes in the temperature, density, or emitting volume in the line forming region. Thus characterizing those changes can potentially lead to a better understanding of the physics that drives these layers. Much of the observational work on temporal fluctuations has centered on the search for evidence of the passage through the transition region of the waves that were thought to heat the corona.

Because of this emphasis on searching for wave motions, the early studies of brightness fluctuations focused on power spectrum analysis of series of brightness observations. Chapman *et al.* (1972), using data from *OSO 7* with a spatial resolution of 10 arc sec × 20 arc sec, found evidence for 300 s intensity fluctuations throughout the transition region and into the corona. Observations at higher resolution and with better statistics have not, however, substantiated these findings.

Using *Skylab* data with a spatial resolution of 5 arc sec × 5 arc sec and improved statistics, Vernazza *et al.* (1975) found no evidence for periodic intensity fluctuations in the transition region and corona. Instead they observed intensity brightenings in quiet solar regions. These brightenings occurred in both network elements and in cell centers, and had greater amplitude in transition region lines than in chromospheric or coronal lines. Based on the equivalent of 20 hours of observing time, Vernazza *et al.* found that the brightenings had an average lifetime, defined as the full width at half-maximum of the emission peak, of 70 s and a mean occurrence time of 330 s. These numbers should be treated as only indicative of the general characteristics of the brightenings. Analyses of power spectra of time series of the brightness fluctuations showed no evidence for periodic behavior in transition region lines. Attempts to search for time lags between fluctuations in different lines suggested that the disturbances were propagating upward. A more detailed analysis, however, was never undertaken, though recently Matheson and Noyes (1990) have reported similar preliminary conclusions from an analysis of Harvard *Skylab* data.

High spatial resolution rocket spectra (Dere *et al.*, 1981), also show brightenings similar to those reported by Vernazza *et al.* (1975). In the C IV resonance lines, the spatial scales for the brightenings are several arc seconds and significant changes can take place in the 20 s intervals between successive exposures. Because of the short exposure times used, however, Dere *et al.* (1981) were only able to observe changes in network emission.

Withbroe (1983) has studied temporal variations in EUV intensities at the limb. Rather than follow individual resolution elements, however, this study looked at the root-mean-square fluctuations over time at each location observed. Withbroe concluded that above 0.03 $R_\odot$ above the limb (about 20,000 km), except for macrospicules, all fluctuations in intensity in the transition region lines were consistent with statistical fluctuations. As with the disk observations, locations showing large fluctuations in the chromospheric lines showed similar fluctuations in all the transition region lines.

Several investigations have also examined the relation between temporal intensity fluctuations and the other parameters of the spectral line, such as the line width and Doppler shift. We will consider those studies in more detail in Chapter 5.

3.4 Relation to Chromospheric Fine Structure

At visible wavelengths the solar chromosphere is observable in optically-thick strong resonance lines, such as Hα and Ca II K. Observations in those lines show not only the same general cell-network pattern seen in the transition region, but also a wealth of small-scale structure.

When observed with a resolution of a few arc seconds, the chromospheric network as seen in the emission cores of the Ca II K line is similar to the transition region network shown in Figure 3.1. Skumanich et al. (1975) performed an analysis of the distribution of Ca II K line emission at 2.4 arc sec resolution. They found that in the middle chromosphere the network as observed in a 1.1 Å band centered on the K line core covers 39% of the surface and contributes 45% of the emission, somewhat smaller than the 60% coverage of the upper chromosphere found by Reeves (1976) at 5 arc sec resolution.

When observed at arc second and better resolution, the chromosphere teems with small-scale structure. These structures have a bewildering variety of names, some of which may refer to the same physical structure as seen at different wavelengths or at the limb or on the disk. Bray and Loughhead (1974) have written a comprehensive review of network fine-scale structure, while Athay (1986b) has produced a briefer overview.

At the limb, spicules dominate the small-scale structure of the chromosphere. They appear as small jet-like features above the limb. A typical spicule diameter is about 1000 km, while they may reach heights of up to 10,000 km above the limb (Beckers, 1972). Time series of observations show that spicules are dynamic. They rise with velocities of about 25 km s^{-1}, reach a peak height, and then either appear to fall back along their upward path or fade from view along their entire length. Determinations of the number of spicules as a function of height above the limb give a scale height for the average decrease in spicule numbers with height above the limb of 1750 km (Beckers, 1972).

This similarity between the scale heights in the emission gradients in UV and EUV lines at the limb and the decrease in spicule numbers, coupled with the extended ragged appearance of the limb in UV and EUV images, has led to the suggestion that some of the EUV transition region emission is from spicules. Early UV and EUV limb observations with low spatial resolution, were interpreted in terms of spicular emission (e.g., Burton *et al.*, 1973; Brueckner and Nicolas, 1973). Attempts to interpret *Skylab* UV and EUV observations in terms of simple empirical models with a spicular component were, however, only partially successful (e.g., Withbroe and Mariska, 1976; Mariska *et al.*, 1978).

Analyses of rocket and *Spacelab 2* filtergrams centered on the C IV resonance lines (e.g., Figure 3.4) have failed to show unambiguous evidence for a one-to-one correspondence at the limb between Hα-emitting spicules and the UV-emitting structures (Cook *et al.*, 1984; Brueckner *et al.*, 1986; Dere *et al.*, 1987). There are, however, individual examples that suggest the connection can be made. Thus, while many investigators seem to have begun to take it for granted that the UV and EUV features seen at the limb are cospatial with or extensions of Hα-emitting spicules, such a conclusion may be premature. Since the word spicule has come to mean any small-scale jet-like feature observed at the solar limb (e.g., Withbroe, 1983), it seems best for the time being always to precede the word with an indication of the wavelength region one is referring to.

On the disk the question of how the fine structure in the UV relates to that seen in Hα and Ca II K is even less clear than it is at the limb. At line center in Hα, the network consists of rosettes, which are clusters of dark fine mottles. Close to the limb these features take on the appearance of bushes. In off-band Hα the bushes appear to be vertical and are usually assumed to be the same structures as the spicules seen at the limb. Between the dark mottles, which form the rosettes, are bright mottles. In active regions the dark mottles tend to be elongated and are referred to as fibrils. The name fibril is also frequently used in the quiet network to refer to dark features that originate in the network and appear to overlie the adjacent supergranule cell. Thus the general picture that emerges is one of a network consisting of nearly vertical structures, which are usually referred to as spicules, and more horizontal structures, which are usually referred to as fibrils.

All this fine-scale structure on the disk is dynamic. Typical lifetimes are about 10 min for mottles and fibrils, while a rosette may last for many hours (Athay, 1986b). Velocity measurements in these structures are difficult. Grossmann-Doerth and von Uexküll (1973) found velocities of 4–8 km s^{-1} in small Hα features near the network.

Within the supergranule cells, bright, time-varying grains seen in the Ca II K line are the main fine-scale chromospheric features (e.g., Liu et al., 1972). The same features are observed in the 1600 Å continuum region, which originates near the temperature minimum at the base of the chromosphere (Cook et al., 1983). Cook and Ewing (1990) have shown that the brightness temperature of these continuum bright points is linearly related to the magnetic field strength.

While it is clear that UV and EUV emission concentrates in the same network regions as the fine structure observed in the chromosphere, the correspondence between structures observed at UV and EUV wavelengths and those seen in Hα and Ca II K remains unclear. High-resolution observations from rockets and *Spacelab 2* in the 1615 Å continuum (which is formed at the temperature minimum), in the C I 1613 Å intercombination lines (which are formed in the chromosphere), and in the C IV resonance lines generally show the same structures (Dere et al., 1987). Figure 3.6 shows an example from the HRTS 6 rocket flight. At 1615 Å and in the C I lines, the average size is about 2000 km (2.8 arc sec), while in the C IV lines the size is about 2450 km (3.4 arc sec). Brueckner and Bartoe (1974) and Dere et al. (1984) have shown that these bright features correlate with the Hα dark mottles. Near the limb, however, the C IV line emission shifts toward the limb in an apparent projection effect. By studying features near the limb, Dere et al. (1986) found that the C I lines formed about 600 km higher in the atmosphere than the continuum, the Fe II 1564 and 1567 Å

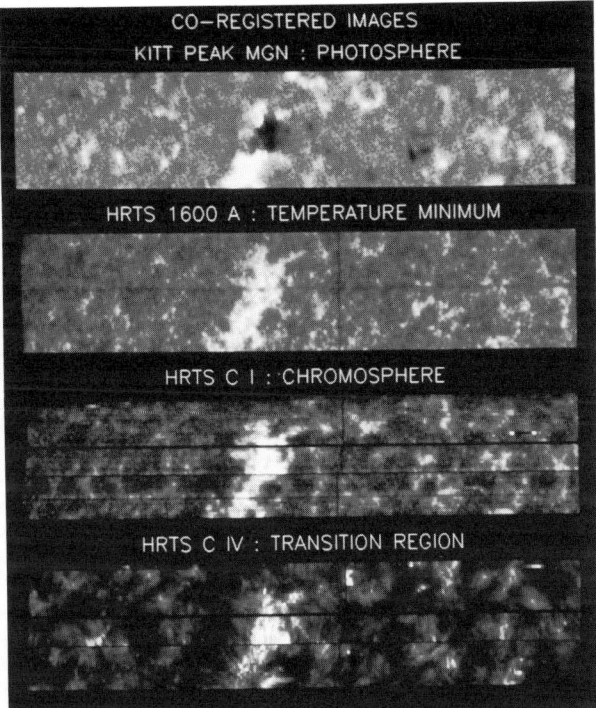

Fig. 3.6. Images of the Sun in the 1615 Å continuum, the 1613 Å C I intercombination lines, and the resonance lines of C IV, along with a coaligned magnetogram from the National Solar Observatory. Each image is 460 arc sec × 107 arc sec (courtesy J. W. Cook, NRL).

lines formed 1000 km above the continuum, and the C IV resonance lines formed 3400 km above the continuum. The average lengths of the features, expressed as the full width at half-maximum intensity, were 1900 km in the continuum, 4400 km in the C I lines, and 4100 km in the Fe II and C IV lines. Thus the features show considerable overlap.

Images of the Sun made through a filter centered at the Lyman α line (Bonnet et al., 1980) have also helped to bridge the gap between visible light chromospheric observations and the more limited transition region data. Figure 3.7 shows an example of one of these images of a polar region. These 1 arc sec resolution images appear to show spicules at the limb, but a one-to-one correspondence with Hα spicules has not been demonstrated. On the disk, the network appears filamentary with some regions showing the bright tops of

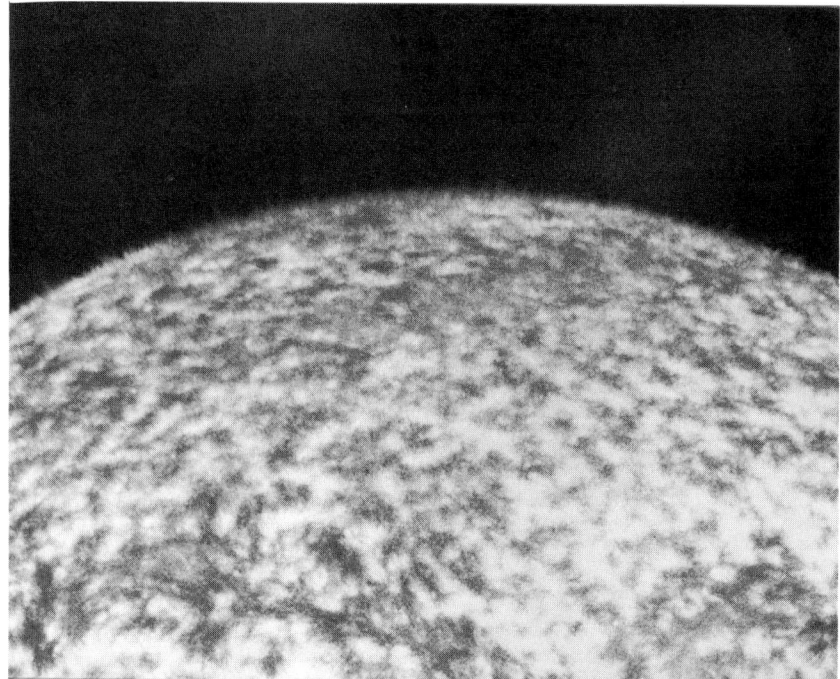

Fig. 3.7. Lyman α filtergram of a polar region (courtesy R. M. Bonnet, ESA-CNRS).

many small loops. Bonnet et al. (1980) also found that large 5–10 arc sec absorption loops appear to cross over brighter emitting regions. In addition, over much of the disk they saw evidence of thin threads, clearly suggesting that large amounts of neutral hydrogen exist in fine-scale structures throughout the height range over which most of the transition region emission is formed and probably into the corona.

All the moderate- and high-resolution data show that the fine-scale structure that dominates much of the emission in Hα and the Ca II K line extends into the transition region. What fraction of the total transition region emission is confined within these small structural elements is uncertain. It is reasonable to assume, however, that nearly all the emission arising in the network is from fine-scale structure. This would suggest, based on the C IV resonance line studies of Dere et al. (1984), that at 10^5 K as much as

85% of the total observed intensity could be concentrated in fine scale structures.

3.5 Relation to the Magnetic Field

Magnetic field observations are generally only available at photospheric levels. These observations show that in the quiet Sun more than 90% of the total magnetic flux clumps into discrete elements, which are primarily in the network (Stenflo, 1989). Moreover, these discrete elements appear to have similar field strengths, between 1000 and 2000 G (e.g., Howard and Stenflo, 1972; Frazier and Stenflo, 1972). The individual magnetic filaments have sizes in the range 100–300 km (Stenflo, 1973), with 100 km being a typical estimate (Stenflo, 1989). Between the magnetic filaments is probably a very small background field. In the quiet Sun this interfilamentary field has a field strength of 10–100 G, is responsible for only 5–10% of the total magnetic flux on the Sun, and also has a characteristic size scale of less than 1 arc sec (Stenflo, 1989).

A magnetic field with a strength of 1000 G exerts tremendous pressure on its environment. The magnetic pressure $B^2/8\pi \approx 4 \times 10^4$ dyn cm^{-2}. Such a strong field can be contained in the photosphere, where the pressure is about 10^5 dyn cm^{-2} at optical depth unity (Vernazza et al., 1981). As the average density model in Figure 1.7 shows, however, the density, and hence the pressure, falls rapidly with increasing height in the solar atmosphere. At the temperature minimum the pressure is down to 10^3 dyn cm^{-2}, and at the 10^5 K level in the transition region it is less than 1 dyn cm^{-2}. Thus a short distance above the optical depth unity point in the photosphere, a magnetic filament must expand. If the nearby magnetic filaments are of the same polarity, then this expansion will end at some height where the fields from the individual elements meet each other and fill the volume. If some of the nearby magnetic filaments have the opposite polarity, then more complex field geometries are possible, with loop-like features being common.

The height in the atmosphere at which the expansion takes place is uncertain. Gabriel (1976) constructed a simple model for the expansion of the field in a unipolar region and obtained a height of about 1500 km for the location where the field expansion occurs. Giovanelli (1980) examined magnetograms obtained in the Mg I b$_2$

line, which is a strong line formed higher in the atmosphere than the lines usually used to obtain magnetograms. He found that these magnetograms showed evidence for field expansion relative to the magnetograms obtained lower in the atmosphere, and deduced a height at which this expansion takes place of 500–600 km above the optical depth unity height. This height, at roughly the temperature minimum level in the Vernazza et al. (1981) model, would place all the observed chromospheric and transition region fine structure in the region that is completely filled with magnetic field.

Just how the fine structure of the chromosphere and transition region fits into this magnetic field picture is unclear. The elongated shapes of the chromospheric and transition region structures suggest that they should correspond to individual bundles of magnetic flux. This view lends additional support to Giovanelli's (1980) observations suggesting that the field expands and develops a significant horizontal component well below the transition region. On the other hand, the observations by Bonnet et al. (1980) of loop-like structures at apparently great heights in the atmosphere, coupled with observations of Lyman continuum absorption apparently reducing the emission of transition region lines (Schmahl and Orrall, 1979; Doschek and Feldman, 1982), suggest that more than a simple expansion of a unipolar field is taking place.

Much of the thinking about the role played in the structuring of the chromosphere and transition region by the magnetic filaments detected at photospheric heights has centered on unipolar fields. Gabriel's (1976) model of the magnetic field expansion, for example, assumes the network elements are unipolar. Much of Reeves' (1976) discussion of the network in EUV lines is in the context of this model. Over many areas of the quiet Sun, however, the magnetic field is not unipolar. Giovanelli (1982) and Dowdy et al. (1986) have argued that most network segments show mixed magnetic polarities, and that this is a general trait of quiet regions. This would imply a magnetic field structure in which some field lines would extend into the corona, but many would close on a size scale corresponding to the separation of the magnetic filaments of opposite polarity—10^4 km or less (Dowdy et al., 1986). Note that even those strands of field that close on a size scale of 10^4 km or less are likely still to undergo some expansion as they arch up into the lower pressure regions above the photospheric layers where they can be confined by the gas pressure.

Figure 3.8 shows an example of a magnetogram of a mixed polarity quiet region and a spectroheliogram obtained in the O VI 1032 Å emission line with the Harvard spectroheliometer on *Skylab*. The O VI image has been computer enhanced to emphasize the small bright features that appear to straddle magnetic dipoles. Dowdy (1990) has argued that these features are small transition region loops, some of which may not achieve coronal temperatures.

If the fine structure observed in the chromosphere and its apparent extension to transition region temperatures are tracing individual bundles of flux, then Hα and UV spicules might be identified with the field lines that extend into the corona, while fibrils might be identified with small-scale closed structures. On the other hand, the fact that in the quiet Sun fibrils tend only to have one end anchored in the network and the other end extending into the supergranule cell without a complete loop-like appearance (Athay, 1986*b*) suggests that they might instead simply lie under the expanding network fields. Additional high-resolution photospheric magnetic field measurements, coupled with observations of both visible and UV fine structure will be required to clarify the relationship between the magnetic field and the fine structure.

3.6 Relation to Coronal Structure

When viewed at a resolution of a few arc seconds in soft X-rays or EUV lines with temperatures of formation over 10^6 K, the dominant feature of the quiet corona is a diffuse vaguely loop-like emission, which is usually referred to as the large-scale structure. This large-scale structure appears to map magnetic field lines in the corona. Because of the low gas pressure there, however, the entire volume of the corona should be filled with magnetic field. Thus the loop-like appearance of the coronal emission suggests that selected sets of magnetic field lines in quiet regions contain more emitting plasma than others. Why this is the case and how the transition region connects to these structures are unknown.

Because both the temperature and density scale heights are much larger in the corona than in the underlying regions of the atmosphere, it is more difficult to uncover the connections of the coro-

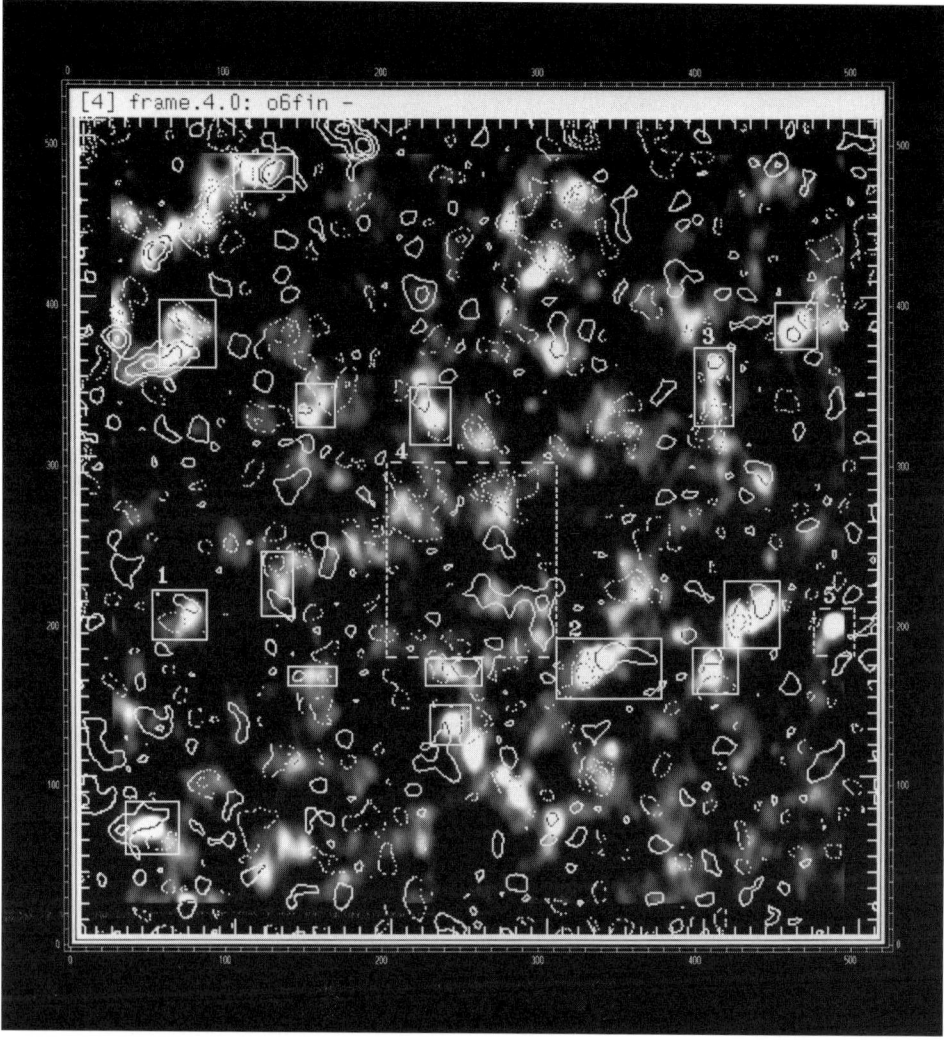

Fig. 3.8. Grey scale representation of a spectroheliogram taken in the 1032 Å emission line of O VI by the Harvard spectroheliometer on *Skylab* on October 2, 1973 at 23:50 UT. It is overlaid with a six-level contour plot of the magnetic flux. The solid contours outline positive flux and the dashed contours are negative. Contour levels are ±10, ±25, and ±45 G. Features outlined with solid-lined boxes show EUV emission over the neutral line separating a small magnetic bipole. The two features enclosed in the dashed-lined box do not show EUV emission over a neutral line (courtesy J. F. Dowdy, Jr., NASA/MSFC).

nal structures to the features seen in the transition region. Often, there are too many overlapping coronal features to locate the exact footpoint of a given loop in the quiet Sun with any precision. Comparisons such as those shown in Figure 3.1 of EUV spectroheliograms obtained near Sun center in the Mg X 625 Å emission line, which is formed at a temperature of roughly 1.5×10^6 K, with simultaneous spectroheliograms obtained in cooler emission lines show that areas of enhanced emission at coronal temperatures always overlay areas of enhanced emission at cooler temperatures (Reeves et al., 1976). Brueckner and Bartoe (1974) found that all the coronal emission they observed in broad band images appeared to be confined within closed loops that originated in UV emission centers. Recent broad band images centered at 173 Å and having a spatial resolution of 1–1.5 arc sec show similar patterns of emission (Walker et al., 1988). Figure 3.9 shows a recent example of a high-resolution EUV broad band image.

While it is probably true that the loops seen in emission lines formed in the corona are the coronal extensions of field lines that do not close back to the surface on the roughly 10,000 km size scale of the network, many details are unclear. How much of the transition region connects to these structures is unknown, as is the role of the fine structure observed in the transition region. For example, it is not clear whether the Hα and UV spicules inject their mass into the loops that we see in emission lines formed at coronal temperatures or are phenomena that follow other bundles of magnetic flux that, though they reach the corona, are not as dense and therefore not as visible as the large-scale structure. Clearly, the global mass and energy balance between the transition region and the overlying corona depends on how these various structures interconnect.

In addition to the large-scale structure, the quiet corona as seen in soft X-rays also exhibits small regions of intense emission. These coronal bright points, which are also observed in coronal holes, are associated with rapidly evolving magnetic bipolar regions (e.g., Harvey et al., 1975 Golub et al., 1977). Sheeley and Golub (1979) found that bright points observed in Fe XV 284 Å spectroheliograms appeared to consist of multiple small loops, each of which evolved on time scales of about 6 min during the roughly 12 hr lifetime of the entire feature.

72 *Emission-Line Intensity Observations*

Fig. 3.9. EUV image of the solar corona at 192.4 Å (Fe XII) obtained on May 13, 1991 with a normal incidence multilayer telescope. Arthur B. C. Walker, Jr. of Stanford University is the Principal Investigator, Richard B. Hoover of the NASA Marshall Space Flight Center and Troy W. Barbee, Jr. of the Lawrence Livermore National Laboratory are Coinvestigators (courtesy NASA, Stanford University, and Lawrence Livermore National Laboratory).

Coronal bright points also exhibit enhanced emission in transition region lines. Habbal and Withbroe (1981) and Habbal *et al.* (1990) found that at transition region temperatures bright points had the same 10–40 arc sec size as coronal bright points, showed evidence for structuring on smaller spatial scales, and showed intensity variability on the 5.5 min temporal resolution of the observations. These brightness fluctuations appear to take place within the individual small structures just as they do in the X-ray bright points. At transition region temperatures, coronal bright points occur in network regions (Habbal *et al.*, 1990). Not all bright features in the transition region network, however, have corresponding coronal emission (Reeves, 1976).

Taken together, the EUV and soft X-ray observations of bright points suggest that they consist of a system of small-scale magnetic loops, each of which is intermittently being heated more or less independently of the others (Habbal and Withbroe, 1981; Habbal *et al.*, 1990). It is still not clear whether coronal bright points are a distinct structural feature or one manifestation of the small-scale structuring of the magnetic field. The bright network features observed in transition region lines, but with no associated coronal emission, may be another manifestation of this small-scale structuring.

Most observations of transition region morphology are at relatively low spatial resolution, 2–5 arc sec. Yet even at that resolution, the observations indicate that there is at least as much structural complexity present at transition region temperatures as ground-based observations already show is present in the chromosphere. In both cases the complex structure is almost certainly intimately tied to the structuring produced by the magnetic field. Here we are further limited both by the inferred subarc second scales of the emerging magnetic field and by the general lack of simultaneous observations of the magnetic field and UV features.

But even with more high spatial resolution observations, a description of the morphology of the transition region can only carry us so far. In the UV the observed emission contains information on both the temperature and the electron density. To make sense of what we see we must separate those two quantities and map the physical conditions in the transition region. That is the subject of the next chapter.

4
Physical Conditions

High spatial resolution images obtained in emission lines formed in the transition region reveal a wealth of detail about the structural morphology of the transition region. To begin to understand the physical processes that are taking place within those structures, however, we must begin to extract physical quantities, such as temperatures, densities, pressures, velocities, and abundance variations from the pictures. Calculations of emission measures are the first step in that process.

4.1 Methods of Calculating Emission Measures

For many emission lines, the simple two-level atomic model outlined in Section 2.3.2 is an excellent approximation for calculating the line flux. Using the Van Regemorter (1962) approximation for the collision strength, we have

$$F = \frac{2.2 \times 10^{-15}}{4\pi R^2} f A_{\rm el} \int g G(T) n_{\rm e}^2 \, dV, \tag{4.1}$$

where the function

$$G(T) = \frac{n_{\rm ion}}{n_{\rm el}} T^{-1/2} \exp\left(\frac{-h\nu}{kT}\right), \tag{4.2}$$

along with the Gaunt factor g, contains all the temperature dependence. Because it contains the relative ion abundance as a factor, the $G(T)$ function for most ions peaks sharply at some temperature $T_{\rm max}$. Taking note of this, Pottasch (1963) assumed that the function had a constant value equal to 0.7 times its maximum value over a temperature range ΔT such that

$$\Delta T \langle G(T) \rangle = \int G(T) \, dT, \qquad (4.3)$$

where

$$\langle G(T) \rangle = 0.70 G(T_{\max}). \qquad (4.4)$$

The Gaunt factor also can be included in this averaging process to define a quantity $\langle g G(T) \rangle$. Often, however, it is simply removed from the integral by assuming it takes the value it has at the peak temperature in the $G(T)$ function. Removing the temperature-dependent factors from the integral, we finally have

$$F = \frac{2.2 \times 10^{-15}}{4\pi R^2} g f A_{\text{el}} \langle G(T) \rangle \int_R n_e^2 \, dV, \qquad (4.5)$$

where the integral is now carried out over the region of formation of the particular emission line. When collision strengths are available, we can write

$$F = \frac{h\nu}{4\pi R^2} \frac{8.63 \times 10^{-6} \Omega}{\omega} 0.8 A_{\text{el}} \langle G(T) \rangle \int_R n_e^2 \, dV. \qquad (4.6)$$

If we assume that the elemental abundances are known, then the observed flux in each emission line determines a single value of $\int_R n_e^2 \, dV$, the emission measure, centered at $T_{\max}$, the peak temperature of the $G(T)$ function, and extending over a temperature interval ΔT defined by equations (4.3) and (4.4).

Most lines that can be approximated using a two-level atomic model have similar values of ΔT. This is not always the case, however, and can lead to two lines formed at the same temperature giving different values of the emission measure because the regions of integration in temperature are different. Jordan and Wilson (1971) proposed selecting a constant logarithmic temperature width $\Delta \log T = \pm 0.15$ dex for all emission lines and then computing a normalization factor such that

$$\frac{\int G(T) \, d\log T}{\Delta \log T} = a G(T_{\max}). \qquad (4.7)$$

The integral is carried out over the entire range of temperatures where $G(T)$ has a significant value, and $\Delta \log T = 0.3$. For resonance lines the value of a is close to 0.7. For many intercombination lines, however, it is 1.0 or larger. This is because the $G(T)$ functions for some of these lines extend over wider temperature domains than those for the resonance lines typically observed in the 1200–1600 Å range. Many recent determinations have used this approach to calculating emission measures (e.g., Mariska, 1980; Raymond and Doyle, 1981a).

Pottasch (1963) and many other early investigators assumed in their derivations of the emission measure that the transition region could be represented by spherically-symmetric shells. This assumption, coupled with the whole-disk observations they used, allowed them to rewrite the volume element in the emission measure integral as

$$dV = \frac{1}{2} 4\pi R_\odot^2 \, dh, \qquad (4.8)$$

where the factor 1/2 accounts for the fact that they only observed half of the atmosphere. Thus the emission measure is often defined as an integral over dh rather than dV. When the line emission is expressed as an intensity (erg cm^{-2} s^{-1} sr^{-1}), the integral is also written over dh (e.g., Raymond and Doyle, 1981a).

These differences can be misleading. It is important to remember physically what is being observed. When we measure a flux or intensity in a given spectral line, we see the sum of all the line photons that are emitted into the line of sight from the detector to the surface of the Sun, and possibly beyond if the line of sight is above the limb. Thus the volume element in equation (4.1) is actually the product of the solid angle subtended by the spectrograph or spectrometer aperture and an element of length along the line of sight. The temperature-dependent function $G(T)$ confines the emission measure to a limited temperature range. In principle, however, the same emission measure could be produced by a low-density plasma occupying a considerable fraction of the line of sight, or a high-density plasma at the same temperature occupying a small fraction of the line of sight. Without further information, the emission measure does not distinguish between these two possibilities.

Despite this ambiguity in interpreting the emission measure, many investigators have used a different definition of the emission measure. In this definition the emission measure integral is rewritten as (e.g., Bruner and McWhirter, 1988)

$$\int_R n_e^2 \, dh = \int n_e^2 \frac{dh}{dT} \, dT. \qquad (4.9)$$

Here the quantity $n_e^2 \mathrm{d}h/\mathrm{d}T$ is referred to as a differential emission measure. If we evaluate the integral in the right hand side of equation (4.9) over a constant logarithmic interval of 0.3 dex, then we see that the emission measure is simply the integral of the differential emission measure over a fixed logarithmic temperature interval. Often the differential emission measure is written as $n_e^2 (\mathrm{d}T/\mathrm{d}h)^{-1}$ to show explicitly the temperature gradient. Occasionally we also see the emission measure integral rewritten as

$$\int_R n_e^2 \, \mathrm{d}h = \int n_e^2 \frac{\mathrm{d}h}{\mathrm{d}\log T} \, \mathrm{d}\log T, \tag{4.10}$$

where the integral on the right side is carried out over $\Delta \log T = 0.30$ as before. The quantity $n_e^2 \mathrm{d}h/\mathrm{d}\log T$ is also sometimes referred to as a differential emission measure. When written in either form, the differential emission measure can easily be used to derive simple models. We will discuss these in Chapter 6.

If multiple distinct structures are present in the line of sight, the integral over volume in equation (4.5) can be quite complex. Craig and Brown (1976) showed that in the general case, with multiple regions at the same temperature in the line of sight, the transformation from a volume element to an integration over temperature must be written as

$$\mathrm{d}V_i = \left(\iint_{S_T} |\nabla T|^{-1} \mathrm{d}S \right)_i \mathrm{d}T, \tag{4.11}$$

where $\mathrm{d}S$ is an element of surface of constant temperature and the index i signifies the ith region in the line of sight in the temperature range T to $T + \Delta T$. Equation (4.1) for the emission line flux then becomes

$$F = \frac{2.2 \times 10^{-15}}{4\pi R^2} f A_{\mathrm{el}} \int g G(T) Q(T) \, \mathrm{d}T, \tag{4.12}$$

where

$$Q(T) = \sum_{i=1}^{N} \left(\iint_{S_T} n_e^2 |\nabla T|^{-1} \mathrm{d}S \right)_i \tag{4.13}$$

is the generalized differential emission measure, and the summation is carried out over all regions in the temperature range T–$T + \Delta T$ in the line of sight.

Of course the structural complexity represented by equation (4.12) is usually not clear in individual line flux observations. Generally one simply ignores the complexity and either makes the formal transformation represented by equation (4.9) or (4.10) after deriving the ordinary volume emission measure (e.g., Jordan and Wilson, 1971), or derives $Q(T)$ directly by analyzing the line flux data using equation (4.12). When this second approach is used, the average value of $Q(T)$ over the range of formation in temperature of the line is often removed from the integral, leaving an integral of $gG(T)$ over temperature, which can be evaluated easily (e.g., Athay, 1966).

Withbroe (1975) has developed an iterative technique for solving for $Q(T)$ when fluxes from several lines are available. Dere and Mason (1981) have, however, criticized his approach. Several investigators have developed methods for finding the differential emission measure as a function of temperature in active regions using X-ray observations (e.g., Sylwester et al., 1980; Siarkowski, 1983). These methods are also applicable to transition region emission line data. Craig and Brown (1986) deal in some detail with the mathematical aspects of this problem.

Much of the controversy over how best to determine the emission measure stems from attempts to go directly from a determination of a quantity such as $Q(T)$ to the atmospheric temperature and density structure. Doing this requires a smooth fit to a set of individual emission measure or differential emission measure determinations that often show considerable scatter. In practice each individual emission line flux measurement can only specify the emission measure over the width of the $G(T)$ function, which is usually about 0.3 dex. If the emission lines used in the analysis are widely separated in temperature, then each line will yield an independent emission measure determination centered at the temperature of formation of the line. If the contribution functions overlap too much, as they often do for X-ray lines from active regions and flares, then the amount of obtainable information is significantly reduced.

In practice this is not usually a problem in the transition region, where there are many emission lines and the $G(T)$ functions peak sharply in temperature. Iterative approaches, such as Withbroe's (1975), may appear not to have enough data to determine a solution for $Q(T)$ to the resolution in temperature claimed. In reality, when the emission lines used in the analysis are widely spaced in temperature, whole sections of the $Q(T)$ function correlate at the level of the width of the $G(T)$ function for the individual lines and

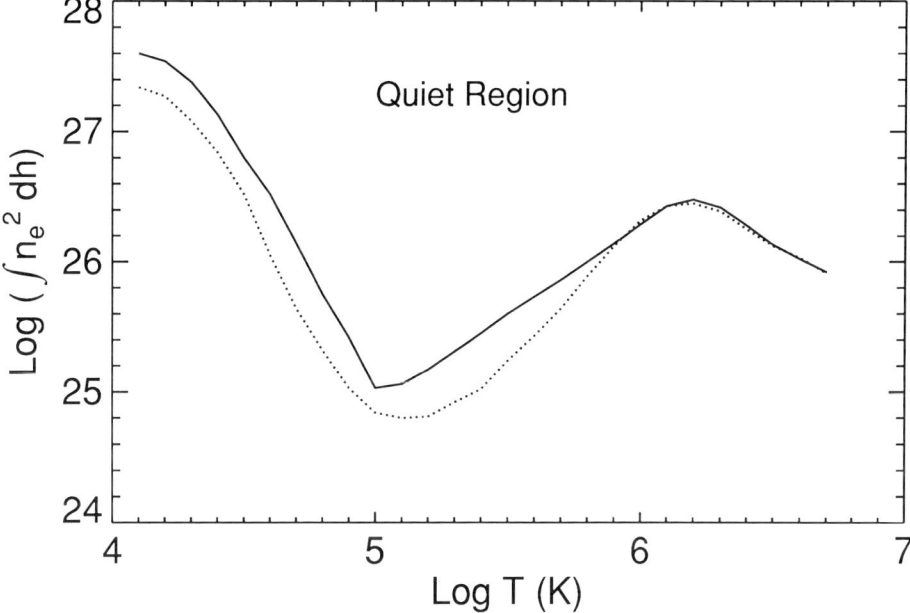

Fig. 4.1. Emission measure distribution for the quiet-Sun network (solid line) and cell center (dotted line) spectra of Vernazza and Reeves (1978). Each value of the emission measure plotted here is integrated over 0.1 in $\log T$ (data courtesy J. Raymond).

the smoothness of the solution at a finer resolution in temperature is misleading.

4.2 Emission Measure Structure of the Transition Region

No matter what technique is used to obtain the emission measure, the resulting distribution as a function of temperature is always remarkably similar. Figure 4.1 shows the emission measure as a function of temperature for quiet-Sun network and cell centers. These curves were obtained by Raymond and Doyle (1981b) based on an analysis of the average network and cell center spectra published by Vernazza and Reeves (1978). This analysis expresses the emission measure as $\int_R n_e^2 \, dh$, where the temperature region is ± 0.05 dex centered on the peak temperature of the ionization fraction for each ion. The emission measures were determined using tables of emissivities calculated by Raymond and Doyle (1981a). Rather than assuming a simple two-level atomic model, these tables were calculated by using computed level populations for the upper level of each transition and the Einstein A-values. They also include an integration through an assumed plane parallel atmosphere at constant pressure.

All emission measure curves derived from quiet-Sun observations display the same general features shown in the figure. There is a minimum in the emission measure near 10^5 K, a power-law distribution with a negative slope at lower temperatures, and a power-law distribution with a positive slope at higher temperatures up to a turnover at coronal temperatures. In average quiet-Sun spectra, the slope of the curve between $\log T = 5.3$ and 6.0 is usually near 1.5. The network and cell center emission measures shown in the figure have slopes of 1.4 and 2.1 respectively. Between $\log T = 4.4$ and 5.0, the cell center emission measure has a slope of -3.3, while the network value is -3.5. Considerable attention has been paid to these slopes in attempts to understand the structure of the transition region. We will consider those analyses in detail in Chapter 6.

The differences between the network and cell center emission measure curves reflect the contrast variations between network and cell centers that we noted earlier. Below 10^5 K, the ratio of the network emission measure to the cell emission measure is about 1.5–2. Near 2×10^5 K, the ratio peaks at a value of 2–3 and then decreases to 1 at coronal temperatures.

These differences between cell center and network emission measures must be viewed with some caution. Vernazza and Reeves (1978) constructed their average spectra by combining individual spectra such that the intensity histograms of several chromospheric and transition region lines were similar to the quiet region histograms obtained by Reeves (1976). As Reeves (1976) pointed out, however, while there is a well-defined average cell intensity distribution, there is no easily defined network average intensity distribution. Network emission is simply what remains after removing the cell center component. Moreover, the recent high-resolution observations clearly suggest that the network contains substantial small-scale structure.

Smooth emission measure curves, such as those shown in Figure 4.1, also tend to gloss over the considerable scatter that exists in the data points from which they were constructed. Many early investigations showed discrepancies at a single temperature as large as a factor of 2 or 3 between lines from ions in different isoelectronic sequences. In recent studies, these discrepancies have been reduced considerably, primarily due to improved atomic physics data allowing corrections for metastable level populations to be included. Raymond and Doyle (1981*b*) claim that the scatter about the smooth lines plotted in Figure 4.1 is 10–20%, and that the slope determinations are accurate to ±0.15. These improvements come, however, at a price. To correct for metastable level populations and for density-dependent effects in the ionization balance, requires an estimate of the electron density in the emitting region and implicitly introduces additional assumptions about the atmospheric structure into the data reduction.

4.3   Abundances

Emission measures from different lines formed at nearly the same temperature often have significantly different values. These differences can arise from many sources, including errors in the line flux measurements, in the instrument calibration, and in the atomic physics constants and ionization balance calculations used in the analysis. In addition, the approximations used in the analysis introduce errors. Analyses that combine data from different experiments also introduce additional errors, since they involve both absolute and relative calibrations and no two experiments observe exactly the same part of the Sun, even when the measurements are simultaneous. Often, however, the emission measures show systematic scatter. If the emission measures from all the lines belonging to the same isoelectronic sequence lie above or below the other emission measures, then the atomic physics needs to be reexamined. If the emission measures from all the lines of a single element fall above or below the other emission measures, then the elemental abundances may be incorrect.

In fact, emission measure analyses are a primary method for determining elemental abundances in the transition region and corona. If we assume that the elemental abundances are unknown, then each emission line flux measurement defines the product of the emission measure and the elemental abundance. Provided that the abundance of each element is constant throughout the transition region and corona, the set of these products for the lines formed from the various ions of the element determines an emission measure curve as a function of temperature. Adjusting the curves for the various elements so that they agree then determines the relative abundances of the elements.

Converting relative abundances to absolute abundances (abundances relative to hydrogen) requires a determination of the electron density in the line forming region. This has been done using radio brightness temperatures, which depend on the temperature and density distribution in the atmosphere (e.g., Pottasch, 1963, 1964), and using measurements of the coronal white light continuum or K-corona, which is produced by the scattering of photospheric light by coronal electrons (e.g., Withbroe, 1970a, 1971). Because of the considerable uncertainty associated with these techniques, absolute transition region and coronal abundances are more uncertain than the relative measurements.

Agreement over the temperature range of the quiet transition region and corona among the emission measure curves from different elements is a strong indication that elemental abundances do not change in those regions of the atmosphere. On the other hand, there is always considerable residual scatter in the individual data points that might be due to abundance variations. Theoretical investigations suggest that in a simple stratified atmosphere, such as that shown in Figure 1.7, there may be mechanisms for producing significant abundance variations between the photosphere and corona (e.g., Delache, 1967; Nakada, 1969; Tworkowski, 1975, 1980; Shine et al., 1975; Meyer and Nussbaumer, 1979). Thermal diffusion, for example, can cause heavier atoms to move upward in the atmosphere.

Observational searches for abundance variations within the transition region itself have produced mixed results. Mariska (1980) analyzed individual spectra obtained at different heights above the solar limb with the NRL slit spectrograph on *Skylab*. To minimize possible variations with temperature, he used lines of C, N, O, and Si that were formed at roughly the same temperature. Forcing all the emission measures to agree with the value determined for the Si IV resonance lines then provided abundances of C, N, and O relative to Si. Because Mariska used emission lines of C to obtain the electron density, only the N and O determinations are reliable. Relative abundances computed for quiet regions, a coronal hole, an active region, and a prominence showed no evidence for significant abundance variations from region to region. They did, however, show that the O and N abundances agreed with the coronal values listed in Table 2.3. This shows that the underabundance of high first ionization potential elements such as O and N relative to their photospheric values discussed in Chapter 2 is already present in the transition region at a temperature of about 60,000 K.

Noci *et al.* (1988) used a standard emission measure analysis technique to examine the average quiet Sun, coronal hole, and active region spectra published by Vernazza and Reeves (1978). They found that the abundances of C, N, O, Ne, Mg, Si, and S in the three regions agreed with those observed at photospheric levels, with the possible exceptions of O and S. At temperatures below 1.6×10^5 K, the O abundance was lower than those obtained in the photosphere, while the S abundance was more than an order of magnitude larger than the photospheric value. The O result relies heavily on the O II line at 833.8 Å, which lies closer to the limit of the hydrogen Lyman continuum at 912 Å than any of the other lines Noci *et al.* considered. In addition, the line blends with stronger emission from O III lines at the same wavelength (Meier *et al.*, 1991). Thus the O result should be viewed with considerable caution. The S result relies on the S II line at 1253.3 Å. In the Harvard EUV spectroheliometer, this line blends with a line of C I at 1253.47 Å and possibly a second C I line at 1254.51 Å (e.g., Feldman *et al.*, 1976a; Sandlin *et al.*, 1986). Thus the S results are probably also unreliable.

Other analyses show some evidence for variations in abundance from region to region on the Sun. Widing and Feldman (1989) analyzed Ne VI and Mg VI multiplets at 400 Å in a variety of active features on the Sun and found that the Ne to Mg ratio varied from 0.1 to 0.2 in open-field structures to 2 to 3 in closed-field structures. In flares the value is close to 4 (Feldman and Widing, 1990). Feldman *et al.* (1990) have also argued that the abundance ratio of C to Si changes from a coronal value of about 4 over a plage to a roughly photospheric value of about 13 over a sunspot.

It is clear from all the coronal and solar wind measurements that there is a depletion of high first ionization potential elements in

84   *Physical Conditions*

the corona relative to the photosphere, with considerable evidence that it is also present in the transition region at temperatures as low as 60,000 K. There also is growing evidence for variations in the transition region and corona from region to region. At present, however, there is no general agreement on a mechanism for producing the abundance variations. Additional studies in a variety of magnetic features may shed some light on possible processes.

4.4 Density Diagnostics

While the emission measure provides us with some idea of the distribution of material in the transition region, our inability to measure directly the volume of the emitting element makes it difficult to interpret. Fortunately, as we outlined in Chapter 2, there are spectroscopic techniques for determining the electron density in the transition region. The large quantities of high spectral resolution data from *Skylab* stimulated a renaissance in the theoretical study of the atomic systems responsible for many of the observed lines. Because of this work, electron density diagnostics are now available for the wide range of physical conditions in the active and quiet solar atmosphere (e.g., Dere and Mason, 1981; Doschek, 1985). For the quiet Sun, emission lines from ions of several isoelectronic sequences are density sensitive. Here we summarize the available data on the important isoelectronic sequences and discuss the derived electron densities.

4.4.1 Beryllium Isoelectronic Sequence

Although line ratio techniques for measuring electron densities have been applied to gaseous nebulae for more than 40 years (e.g., Aller *et al.*, 1949), their application to UV spectra and the solar transition region is more recent. The first density-sensitive line ratios proposed were for ions in the Be I isoelectronic sequence (Munro *et al.*, 1971; Jordan, 1971). Table 4.1 lists the important emission lines of Be-like ions in the solar UV and EUV spectrum.

Figure 4.2 shows a partial term diagram for the C III ion. The Be-like ions have a $2s2p$ first excited configuration that results in a ^1P and a ^3P term. The ^1P level is excited from the ^1S ground state and produces the 977 Å allowed line. The ^3P term is also excited from the ground state, but is metastable. It radiatively couples

Table 4.1. *Important emission lines of Be-like ions.*

		Wavelength (Å)	
Transition	C III	N IV	O V
$2s^2\ {}^1S_0 - 2s2p\ {}^1P^o_1$	977.03	765.15	629.71
$-2s2p\ {}^3P^o_1$	1908.73	1486.50	1218.35
$2s2p\ {}^3P^o - 2p^2\ {}^3P$	1175.7	923.1	760.4
${}^3P^o_0 - {}^3P_1$	1175.26	922.52	759.44
${}^3P^o_1 - {}^3P_0$	1175.99	923.68	761.13
${}^3P^o_1 - {}^3P_1$	1175.59	923.06	760.23
${}^3P^o_1 - {}^3P_2$	1174.93	921.99	758.68
${}^3P^o_2 - {}^3P_1$	1176.37	924.28	762.00
${}^3P^o_2 - {}^3P_2$	1175.71	923.22	760.44
$2s2p\ {}^1P_1 - 2p^2\ {}^1S_0$	1247.38	955.34	774.52
$-2p^2\ {}^1D_2$	2296.87	1718.55	1371.29

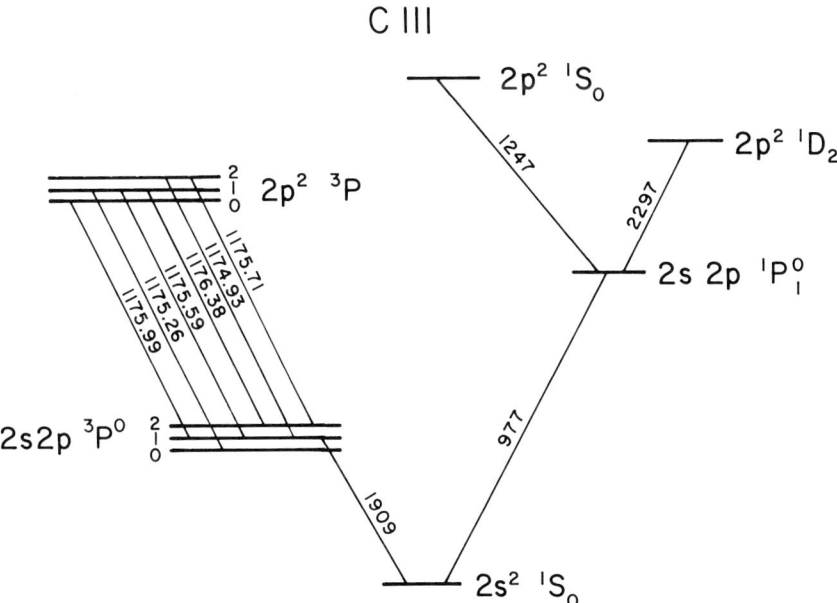

Fig. 4.2. Partial term diagram for C III.

to the ground state through the $2s^2\ {}^1S_0$–$2s2p\ {}^3P_1$ electric-dipole transition, which produces the 1909 Å intersystem line, and the $2s^2\ {}^1S_0$–$2s2p\ {}^3P_2$ magnetic quadrupole transition, which produces the 1907 Å forbidden line. For nebular densities the ratio of the 1907 Å line to the 1909 Å line is density sensitive and behaves like the simple three-level case we discussed in Chapter 2.

At transition region densities different density-sensitive line ratios become important. Since the $2s2p\ {}^3P$ term is metastable, lines collisionally excited from it will be density sensitive in the manner of the four-level case we discussed in Chapter 2. Thus the ratio of lines originating in the $2s2p\ {}^3P$ term to the allowed 977 Å transition will be density sensitive. The useful range of this diagnostic depends on the atomic model used. Other line ratios within the C III system are also density sensitive for typical solar electron densities. These include $I(1176\ \text{Å})/I(1247\ \text{Å})$, $I(1247\ \text{Å})/I(1909\ \text{Å})$, and $I(1176\ \text{Å})/I(1909\ \text{Å})$.

Detailed atomic data for electron collision rates and the resulting level populations for C III have been published by Berrington et al. (1977), Dufton et al. (1978), Keenan et al. (1984), Keenan and Berrington (1985), and Berrington et al. (1985). Using the Dufton et al. (1978) collision strengths and transition probabilities for allowed and forbidden lines from Mühlethaler and Nussbaumer (1976), Cook and Nicolas (1979) investigated the line ratios outlined above. They found that all the ratios gave consistent results except those involving the 1176 Å lines originating in the $2p^2\ {}^3P$ term. It appeared that these differences were the result of the 1176 Å lines being about a factor of 5 lower in intensity than the expected optically thin value. This may have been due to an absolute calibration error in the NRL *Skylab* spectrograph. The fact that Cook and Nicolas were able to determine a consistent density from the $I(1176\ \text{Å})/I(977\ \text{Å})$ line ratio data published by Dupree et al. (1976) partially supports this suggestion. Doyle and McWhirter (1980) have shown that at the limb there is a small amount of opacity in the strongest line of the 1176 Å multiplet, while at Sun center there is none. The 977 Å resonance line will have considerable opacity at the limb, but is effectively thin at disk center. Densities determined using this diagnostic are, however, subject to considerable uncertainty, since the theoretical line ratio varies by only a factor of 2 for a change in the density of a factor of 20.

Using quiet-Sun limb data within 4 arc sec of the white-light limb, Cook and Nicolas (1979) found an average electron density of 2.4×10^{10} cm^{-3} at a temperature of 56,000 K from the $I(1247$ Å$)/I(1909$ Å$)$ diagnostic. The standard deviation of that average is 1.3×10^{10} cm^{-3}. Their value of the electron density from the Dupree et al. (1976) quiet-Sun data for $I(1176$ Å$)/I(977$ Å$)$ was 5×10^9 cm^{-3}. This is close to the value of 4.6×10^9 cm^{-3} obtained by Dupree et al. with empirically adjusted atomic parameters. Using improved atomic physics calculations, Keenan and Berrington (1985) obtained an electron density of 1.9×10^9 cm^{-3} from this diagnostic.

Of the density-sensitive ratios in the C III spectrum, the $I(1247$ Å$)/I(1909$ Å$)$ ratio is the most reliable. It is, however, only useful for observations obtained near the solar limb. The presence of a strong continuum hampers observations of the 1909 Å emission line on the disk. In addition, the 1247 Å line is intrinsically weak and blends with a nearby unidentified neutral line, which is not present above the limb. Figure 4.3 shows the calculated $I(1247$ Å$)/I(1909$ Å$)$ ratio as a function of electron density.

Uncertainty about the role of proton collisions introduces an additional source of error in the theoretical line ratios. These collisions redistribute the populations within the ^3P term and thus affect emissivities in the 1176 Å lines and in the 1909 Å line. Dufton et al. (1978) considered cases in which the proton collision rates were zero, equal to the electron collision rates, and a factor of 10^6 greater than the electron collision rates. They found that for electron densities greater than about 10^9 cm^{-3}, electron collisions were rapid enough to mix the populations of the fine structure levels. Their estimate, however, was for an assumed temperature of 70,000 K. For lower temperatures, the choice of proton collision rates is more important.

All the C III theoretical line ratios also suffer to some extent from a dependence on temperature. For example the theoretical $I(1247$ Å$)/I(1909$ Å$)$ ratio plotted in the figure increases from a value of 0.02 at 56,000 K to 0.1 at 1.26×10^5 K. Thus it is important either to be confident that the observed lines are produced at the peak temperature of formation of the ion, or to verify (using emission measures and semiempirical models) that the density

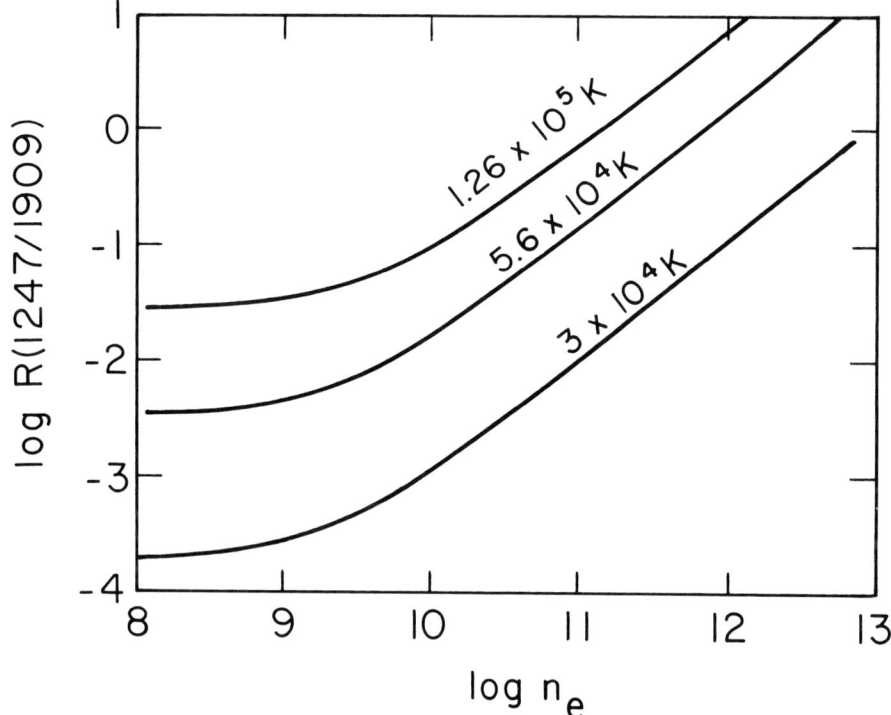

Fig. 4.3. The ratio $I(1247\text{ Å})/I(1909\text{ Å})$ as a function of electron density at three temperatures of formation for the lines (after Cook and Nicolas, 1979).

estimate is consistent with the derived physical conditions in the region.

Other ions in the Be I isoelectronic sequence also have density-sensitive line ratios. For O V, Berrington et al. (1977), Dufton et al. (1978), Keenan et al. (1984), and Berrington et al. (1985) have published collision rates and level populations. Widing et al. (1982) and Doyle et al. (1983) have extended those calculations to include the $n = 3$ states. Spontaneous radiative decay rates have been published by Hibbert (1980) and Nussbaumer and Storey (1979b). As with C III, the ratio $I(^3P^e \to {}^3P^o)/I({}^1P^o \to {}^1S^e) = I(760\text{ Å})/I(630\text{ Å})$ is density sensitive. The observed quiet-Sun value of this ratio is, however, near or at the low-density limit, indicating that the electron density is below about 10^{10} cm^{-3}. Only when the electron density exceeds about 10^{11} cm^{-3} does the line ratio become a reliable diagnostic. Doyle et al. (1983) suggest that the $2s2p\,^3P^o_1 \to 2p^2\,^3P^e_0$ line at 761.1 Å is very sensitive to density in the range 10^9–10^{12} cm^{-3}. This line is not, however, resolved in currently available data. Widing et al. (1982) have investigated $n = 3$ to $n = 2$ transitions in the wavelength interval from 190–200 Å, and found some potentially useful diagnostics for the transition regions of flares. Thus with currently available data, none of the O v line ratios are useful quiet-Sun density diagnostics.

For N iv Dufton et al. (1979) Keenan et al. (1984), and Keenan et al. (1986a) have interpolated collision rates using calculated C iii and O v rates and published level populations. Spontaneous decay rates have been published by Hibbert (1974), Glass and Hibbert (1978), and Nussbaumer and Storey (1979b). At quiet-Sun densities, the ratio $I(^3P^e \to \,^3P^o)/I(^1P^o \to \,^1S^e) = I(923.1\text{ Å})/I(765.15\text{ Å})$ is density sensitive. Unfortunately, both lines suffer from blends in the currently available data, making evaluation of the quiet-Sun density unreliable. After correcting for blends, Dufton et al. (1979) estimated a quiet-Sun electron density of approximately 1.5×10^{10} cm^{-3} from this ratio at a temperature of 1.26×10^5 K. This estimate is based on off-limb data from the Harvard instrument on *Skylab*. Using improved excitation rates, Keenan (1990) estimated an electron density of 1.3×10^{10} cm^{-3}. The ratio $I(^1D^e \to \,^1P^o)/I(^3P^o_1 \to \,^1S^e) = I(1718.55\text{ Å})/I(1486.5\text{ Å})$ is sensitive to both density and temperature. For quiet-Sun densities it is near the low-density limit and thus is of limited usefulness as a density diagnostic.

4.4.2 Magnesium Isoelectronic Sequence

In the Be i isoelectronic sequence, the last two electrons are in the L shell and the ground level is $2s^2\,^1S_0$. In the Mg i isoelectronic sequence, electrons fill the L shell and the last two electrons are in the M shell, leading to a ground level of $3s^2\,^1S_0$, and an energy level structure that is similar to the Be-like ions. Thus we expect there to be similar density-sensitive line ratios in the two sequences.

For quiet-Sun densities, the key ion in the Mg i sequence is Si iii. Figure 4.4 shows a partial term diagram for Si iii with the key UV transitions indicated. Nicolas et al. (1979) and Dufton et al. (1983) have investigated the density sensitivity of the emission lines from Si iii observed in the solar transition region. The Dufton et

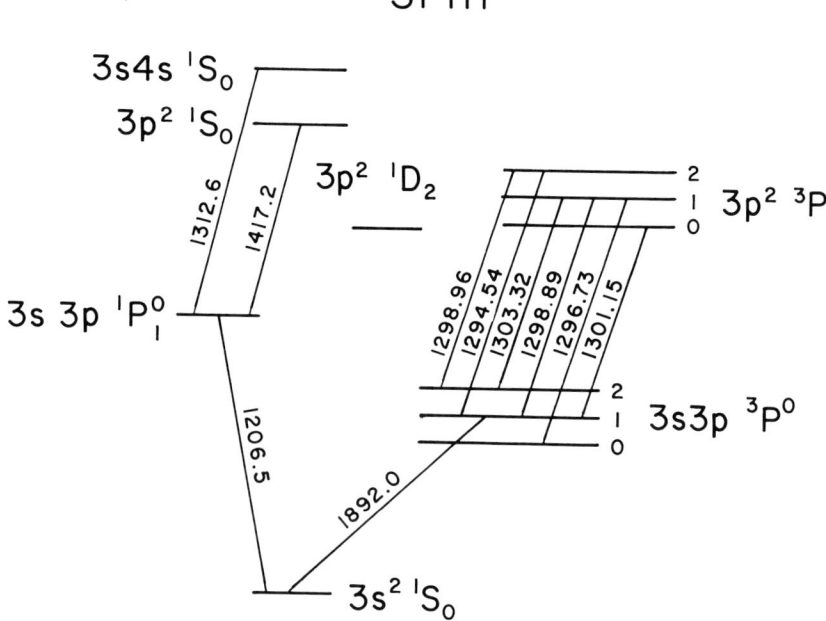

Fig. 4.4. Partial term diagram for Si III.

al. calculations include the effects of resonances in the collision strengths at low energies and are thus more accurate. In the UV portion of the spectrum, there are four density-sensitive line ratios:

$$R_1 = I(1296\ \text{Å})/I(1892\ \text{Å}), \tag{4.14}$$
$$R_2 = I(1301\ \text{Å})/I(1313\ \text{Å}), \tag{4.15}$$
$$R_3 = I(1301\ \text{Å})/I(1296\ \text{Å}), \tag{4.16}$$

and

$$R_4 = I(1301\ \text{Å})/I(1303\ \text{Å}). \tag{4.17}$$

Figures 4.5, 4.6, and 4.7 show the density dependence of these ratios. The ratios R_1 and R_2 show significant temperature dependence. Moreover, the large difference in wavelength between the two lines in R_1 leads to additional uncertainty associated with errors in instrumental calibration. All the ratios do not reach their full density sensitivity until the electron density is at or above 10^{10} cm^{-3}, so typical quiet-Sun densities are near, but not at, the low-density limit. Despite these problems, these ratios are among the best density diagnostics for the quiet solar transition region.

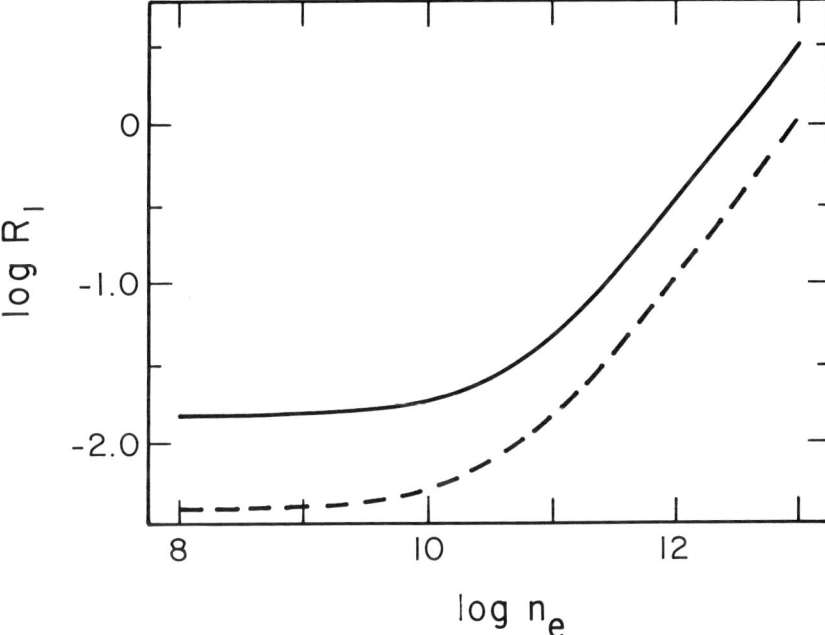

Fig. 4.5. The ratio $R_1 = I(1296\ \text{Å})/I(1892\ \text{Å})$ as a function of electron density. The solid line is for a temperature of 50,000 K and the dashed line is for a temperature of 32,000 K (after Dufton et al. 1983).

Dufton et al. (1983) used quiet-Sun spectra taken between 2 and 6 arc sec above the limb and obtained values for the R_3 and R_4 ratios of 0.40 and 0.31, respectively. These implied electron densities at a temperature of 32,000 K of 1.58×10^{10} and 1.26×10^{10} cm^{-3}, respectively. Keenan et al. (1989) examined HRTS data from *Spacelab 2* and found solar disk quiet-Sun values for the R_3 and R_4 ratios of 0.55 and 0.42, leading to an electron density estimate of 2.51×10^{10} cm^{-3}. Other limb observations give values for the ratios that are closer to those determined by Dufton et al. (1983). For example, Nicolas et al. (1979) found a value of 0.39 for R_3 in a HRTS rocket spectrum, and Kjeldseth Moe and Nicolas (1977) obtained average values for R_3 and R_4 between 2 and 6 arc sec above the limb of 0.38 and 0.33.

Dufton et al. (1983) also determined that the ratio R_2 yielded consistent electron densities if they used a temperature of line formation of 50,000 K. On the other hand, Keenan et al. (1989) have

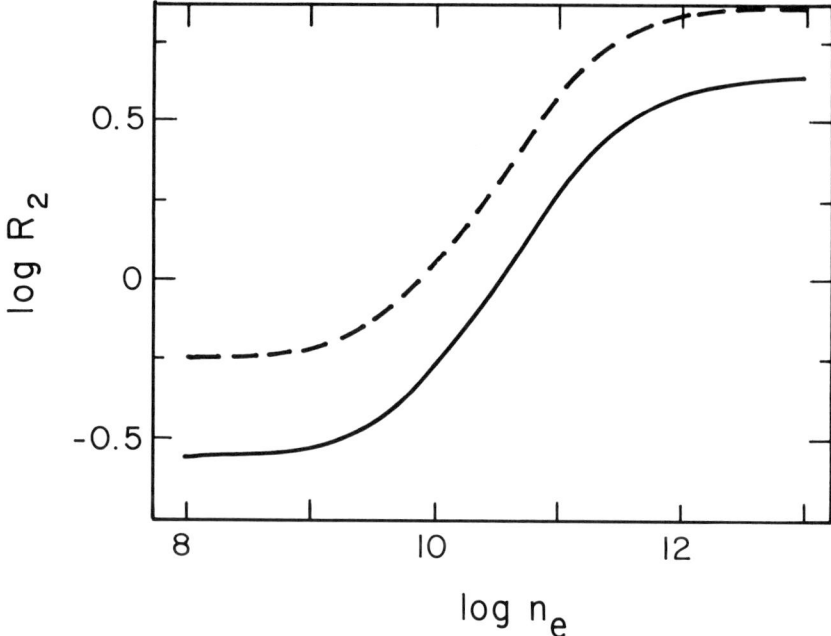

Fig. 4.6. The ratio $R_2 = I(1301 \text{ Å})/I(1313 \text{ Å})$ as a function of electron density. The solid line is for a temperature of 50,000 K and the dashed line is for a temperature of 32,000 K (after Dufton et al. 1983).

suggested that this ratio may be affected by the possible presence of a non-Maxwellian electron distribution function. Thus the R_3 and R_4 ratios are the most useful.

The ratio $I(1298 \text{ Å})/I(1206 \text{ Å})$ is also density sensitive. Like the corresponding ratio in the C III ion, however, the range of sensitivity is small, about a factor of 5 (Nicolas et al., 1979). In addition the large intensity difference between the strong resonance line and the weaker intersystem lines makes them difficult to measure in a single observation. Finally, the resonance line has an optical depth at Sun center of about 7 (Dufton et al., 1983) violating the optically thin assumption.

4.4.3 Boron Isoelectronic Sequence

Ions in the B I isoelectronic sequence also have emission line ratios that are sensitive to electron density. Figure 4.8 shows a partial energy level diagram for the B-like ion Mg VIII. In many ways the atomic structure is similar to the Be-like ions. Transitions

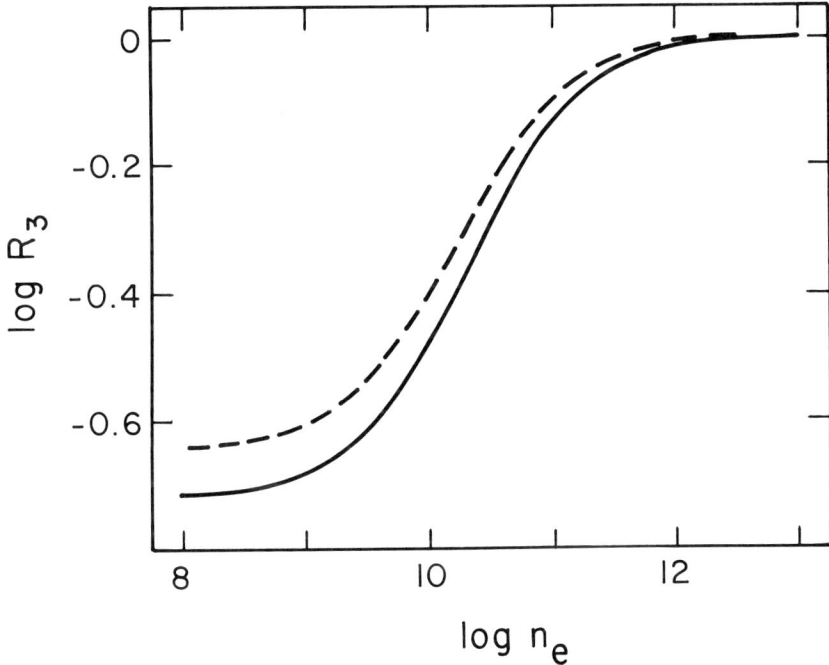

Fig. 4.7. The ratio $R_3 = I(1301\ \text{Å})/I(1296\ \text{Å})$ as a function of electron density. The solid line is for a temperature of 50,000 K and the dashed line is for a temperature of 32,000 K. The ratio

$$R_4 = I(1301\ \text{Å})/I(1303\ \text{Å})$$

has the same density dependence, with $\log(R_4/R_3) = 0.09$ (after Dufton et al. 1983).

from the $2p^2\ ^2D$ multiplet to the $2p^2\ ^2P$ ground state are allowed, and the $2p^2\ ^4P$ multiplet is metastable. Thus collisionally excited transitions from levels in the $2p^2\ ^4P$ term will be density sensitive as we discussed for the four-level case in Chapter 2. For example, in the N III ion, the ratio of the intensity of the $2p^2\ ^4S \to 2p^2\ ^4P$ line at 772.0 Å to the $2p^2\ ^2D_{3/2} \to 2p\ ^2P_{1/2}$ line at 989 Å is sensitive to density up to an electron density of 10^{11} cm^{-3}.

Vernazza and Mason (1978) have investigated the density sensitivity of the B-like ions N III, O IV, Ne VI, Mg VIII, and Si X. They concentrated on the 300–1350 Å wavelength range of the Harvard College Observatory spectrometer flown on *Skylab*, and found many useful ratios, even at the 1 Å resolution of the instrument. For N III, O IV, and Ne VI, the ratio of the $2p^3\ ^4S \to 2p^2\ ^4P$ transitions to one

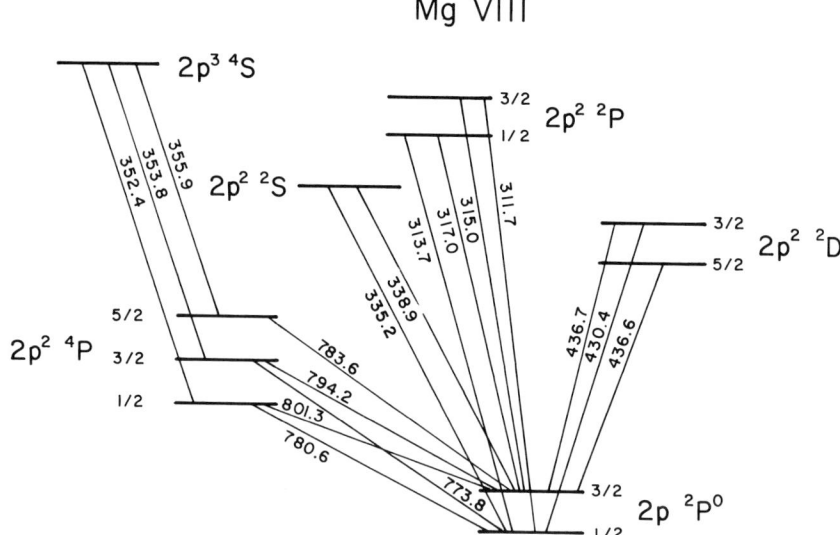

Fig. 4.8. Partial term diagram for Mg VIII.

of the allowed $2p^2\ {}^2D \to 2p\ {}^2P$ transitions is density sensitive at quiet-Sun densities. Ratios of this type investigated by Vernazza and Mason include $I(772\text{ Å})/I(989\text{ Å})$ in N III, $I(625\text{ Å})/I(790\text{ Å})$ in O IV, and $I(454\text{ Å})/I(562\text{ Å})$ in Ne VI.

The N III $I(772.4\text{ Å})/I(989.8\text{ Å})$ ratio is theoretically an excellent density diagnostic at densities below about 10^{11} cm^{-3}. Unfortunately, at the resolution of the Harvard spectrometer on *Skylab*, the $2p^3\ {}^4S \to 2s2p^2\ {}^4P$ transitions blend with a strong Ne VIII line at 770 Å, making their intensities difficult to evaluate. Accounting for this blend, Vernazza and Mason determined a quiet-Sun value for this ratio of 0.046 and obtained an electron density of 4.0×10^9 cm^{-3} at a temperature of 80,000 K.

The O IV $I(625\text{ Å})/I(790.1\text{ Å})$ ratio is also theoretically an excellent density diagnostic. For an expected quiet-Sun density of about 10^{10} cm^{-3}, the theoretical value of the ratio should be about 0.01. The $2s2p^2\ {}^2D \to 2s2p\ {}^2P$ transition at 790.1 Å is observed in the Harvard *Skylab* data with a count rate of 45.70 counts per 0.041 s, corresponding to an intensity of 83.15 erg cm^{-2} s^{-1} sr^{-1}. This means that the expected intensity of the 625 Å lines from the $2p^3\ {}^4S \to 2s2p^2\ {}^4P$ transitions should be ∼0.83 erg cm^{-2} s^{-1} sr^{-1}, leading to a count rate of only about 0.46 counts per 0.041 s, if we ignore calibration differences between the two wavelengths. These transitions are not only intrinsically weak, they also blend with the much stronger Mg X coronal line at 625.2 Å. Thus, while the O IV diagnostic looks promising, it is not yet possible to apply it because of both wavelength resolution and sensitivity problems.

For higher atomic weight ions in the isoelectronic sequence, the density at which the line ratio saturates increases. This leads to smaller values for the theoretical ratio at typical quiet-Sun densities. Thus for Ne VI, the $I(454\text{ Å})/I(562\text{ Å})$ theoretical ratio should have a value of only about 0.0005 for an electron density of 10^{11} cm^{-3}. In the Harvard *Skylab* data, the $2s2p^2\,^2D \rightarrow 2s2p\,^2P$ transitions at 562.6 and 558.5 Å may have been detected. Clearly, however, extremely good sensitivity will be required to detect the $2p^3\,^4S \rightarrow 2s^22p\,^4P$ transitions at 451.8, 452.8, and 454.1 Å. Thus this diagnostic is not promising for quiet-Sun densities.

For heavier ions in the isoelectronic sequence, other line ratios are density sensitive at quiet-Sun densities. In Mg VIII the ratios $I(783\text{ Å})/I(430\text{ Å})$ and $I(794\text{ Å})/I(430\text{ Å})$ are mildly density sensitive. The 783 and 794 Å lines originate in the metastable $2p^2\,^4P$ term, while the 430 Å line originates in the $2p^2\,^2D$ term. These ratios only vary by a small amount, however, and are no longer density sensitive above 10^{11} cm^{-3}. Vernazza and Mason quote a quiet-Sun value for the $I(794\text{ Å})/I(430\text{ Å})$ ratio of 0.035, implying an electron density of 1.0×10^8 cm^{-3} at a temperature of 10^6 K. The analogous ratios in the Si X ion are $I(638\text{ Å})/I(347\text{ Å})$ and $I(653\text{ Å})/I(347\text{ Å})$. Again, however, the ratios only vary by a small amount.

Beginning with Mg VIII, the ratio of lines between the $2p^2\,^2D$ excited state and the $2p\,^2P$ ground state begins to show density sensitivity at quiet-Sun densities. At low densities electrons only populate the $2s^22p\,^2P_{1/2}$ ground level. For higher densities the $2s^22p\,^2P_{1/2}$ and $2s^22p\,^2P_{3/2}$ populations are in Boltzmann equilibrium. Thus in the manner of the four-level case discussed in Chapter 2, the ratio of the emission lines preferentially excited from the two levels will be density sensitive. In Mg VIII the relevant ratio is $I(430\text{ Å})/I(436\text{ Å})$. This ratio is sensitive up to a density of about 10^9 cm^{-3}. Vernazza and Mason found an observed quiet-Sun value of 0.88 for this ratio, implying an electron density of 4×10^8 cm^{-3} at a temperature of 10^6 K. In Si X the analogous ratio is $I(347\text{ Å})/I(355\text{ Å})$. Vernazza and Mason found an observed quiet-Sun value for this ratio of 2.70, implying an electron density of 2×10^8 cm^{-3} at 1.5×10^6 K. In S XII, the analogous ratio is $I(288\text{ Å})/I(299\text{ Å})$, for which there are no accurate quiet-Sun observations.

At longer wavelengths, transitions in N III and O IV between the metastable $2p^2\ ^4P_J$ levels and the $2p^2\ ^2P_{J'}$ ground levels are density sensitive. As the electron density increases, the upper levels move toward Boltzmann equilibrium, leading to density sensitivity. In N III the key lines are the $J = 5/2$ to $J' = 3/2$ line at 1749.7 Å, the $J = 1/2$ to $J' = 3/2$ line at 1754.0 Å, and the $J = 3/2$ to $J' = 3/2$ line at 1752.2 Å. The key ratios are $I(1754.0\text{ Å})/I(1749.7\text{ Å})$, and $I(1752.2\text{ Å})/I(1749.7\text{ Å})$. Nussbaumer and Storey (1979a) have investigated these ratios and found density sensitivity in the 10^9–10^{11} cm^{-3} range, which should make them excellent for the quiet Sun and coronal holes. Unfortunately, the ratios only vary by about a factor of 2.5 over this density range, suggesting that small errors in measuring line intensities will result in sizable uncertainties in the derived density.

The 1754.0 and 1749.7 Å lines have been measured in the quiet Sun by Doschek et al. (1976b). On the disk the 1754.0 Å line is blended. Above the limb, the average value of the line ratio is 0.23, which yields an electron density of about 2.5×10^9 cm^{-3} at a temperature of 30,000 K. This value is lower than those determined with some other diagnostics, suggesting that either blends may be a factor or that more detailed atomic physics calculations may be necessary before it can be applied. Feldman and Doschek (1979) combined a larger set of data from the NRL *Skylab* spectrograph to determine an average ratio for both coronal holes and the quiet Sun, obtaining an electron density of 2×10^9 cm^{-3}, again much lower than accepted values. They pointed out, however, that changes of 10% in some of the key transition probabilities would significantly increase the derived density.

In O IV the important lines are the $J = 5/2$ to $J' = 3/2$ line at 1401.2 Å, the $J = 1/2$ to $J' = 3/2$ line at 1407.4 Å, and the $J = 3/2$ to $J' = 3/2$ line at 1404.8 Å. Nussbaumer and Storey (1982) have investigated the ratios $I(1407.4\text{ Å})/I(1401.2\text{ Å})$ and $I(1404.8\text{ Å})/I(1401.2\text{ Å})$ and again found density sensitivity for electron densities between 10^9 and 10^{11} cm^{-3}. Like the N III diagnostic, however, the ratios only vary by about a factor of 2.5 over the range of density sensitivity. The average value for the first ratio measured near the limb by Doschek et al. (1976b) is 0.16, which falls near or below the low-density limit for the theoretical diagnostic, suggesting that additional theoretical work is required before this can be a useful diagnostic. As with the similar N III density diagnostics, small changes in key transition probabilities could result in significant changes in the theoretical ratios (Feldman and Doschek 1979).

Dere et al. (1982) adjusted the Flower and Nussbaumer (1975b) collision strength calculations to account for resonances and computed the ratios $I(1401.2$ Å$)/I(1404.8$ Å$)$ and $I(1399.8$ Å$+1407.4$ Å$)/I(1404.8$ Å$)$. Using these ratios and NRL HRTS rocket data for a quiet limb, they derived an average electron density of 7.7×10^9 cm^{-3} at a temperature of 1.3×10^5 K.

Theoretical calculations for the B-like ions are not as complete as those for the Be-like ions. This is due in large part to the lack of high-resolution observations at the wavelengths of many of the emission lines. Thus in some cases, theoretical line ratios disagree with the limited observations in the high-density limit, leading to empirical adjustments in some collision rates. For example, Vernazza and Mason (1978) adjusted some collision rates in their calculations of line ratios for O IV, Ne VI, Mg VIII, and S XII by 15% to achieve the observed high-density limit for the $I(783$ Å$)/I(430$ Å$)$ ratio in Mg VIII. An additional adjustment to the data was required for them to obtain satisfactory densities for some of the Si X lines they observed. Further progress on density sensitivity in these ions must await improved observations in the 100–1000 Å wavelength range.

4.4.4 Aluminum Isoelectronic Sequence

The Al I isoelectronic sequence is similar to the B I isoelectronic sequence in the same way that the Mg I isoelectronic sequence is similar to the Be I isoelectronic sequence. At quiet-Sun densities, lines from S IV are near the low-density limit, but have some sensitivity. Atomic physics calculations for S IV have been performed by Bhatia et al. (1980), Bhadra and Henry (1980), and Dufton et al. (1982). The Dufton et al. calculations, which expand on earlier calculations by Dufton and Kingston (1980), include the effects of resonances at low energies in the collision strengths and should be more accurate than the earlier calculations. The decay rates calculated by Dufton et al. also differ from earlier determinations.

In the UV, the diagnostically interesting emission lines connect the metastable $^4P_j^e$ levels with the $^2P_{j'}^o$ levels, just as in the B sequence. The important lines are the $J = 3/2$ to $J' = 1/2$ line at 1416.9 Å, the $J = 5/2$ to $J' = 3/2$ line at 1406.0 Å, and the $J = 1/2$ to $J' = 3/2$ line at 1423.9 Å. Dufton et al. investigated the two ratios $R_1 = I(1416.9 \text{ Å})/I(1406.0 \text{ Å})$ and $R_2 = I(1423.9 \text{ Å})/I(1416.9 \text{ Å})$. Using several Skylab quiet-Sun observations between 2 and 6 arc sec above the limb, they found average values of $R_1 = 0.62$ and $R_2 = 0.26$. These values implied values of $\log n_e$ of 10.3 and 11.2, respectively at a temperature of 90,000 K. The 1416.9 Å line is quite weak in the quiet Sun and the ratios are near the low-density limit. Thus the difference between these values is probably more a reflection of the difficulty of applying these ratios to quiet-Sun data than an indication that the atomic data are suspect. Densities determined for active regions and flares are more consistent between the two ratios (Dufton et al., 1982).

Dufton and Kingston (1985) have also investigated density-sensitive line ratios in the Al-like ion Si II. At quiet-Sun densities the ratio $I(2334.61 \text{ Å})/I(2350.17 \text{ Å})$ is near the high-density limit, but still density sensitive. The 2334.61 Å intercombination line includes two components at 2334.40 Å ($^4P_{1/2} \rightarrow {}^2P_{1/2}$) and 2334.61 Å ($^4P_{5/2} \rightarrow {}^2P_{3/2}$), which blend in currently available spectra. In addition, because of the strong continuum at long wavelengths, only observations above the limb provide useful spectra for these lines. Using NRL Skylab data and taking this blend into account, Dufton and Kingston determined a value of 4.2 for this ratio. This results in an electron density of 10^{11} cm^{-3} at a temperature of 10^4 K.

4.4.5 Carbon Isoelectronic Sequence

Lines from the O III ion in the carbon isoelectronic sequence have been used for many years as a temperature diagnostic in gaseous nebulae (e.g., Osterbrock, 1974). Both O III and ions of heavier elements in the sequence have density-sensitive lines. At quiet-Sun densities lines of Mg VII, Si IX, and S XI are density sen-

Density Diagnostics

sitive. These lines fall in the 170–400 Å spectral range. Mason and Bhatia (1978) and Aggarwal (1984) have computed atomic data for these ions and Dere et al. (1979) have discussed the density-sensitive line ratios that are available and tabulated flare observations that exhibit the lines. The primary source of observations in this wavelength range is the NRL slitless spectrograph flown on *Skylab*. Because it was slitless, a high-intensity small-scale feature, such as a flare, provides the best means of identifying emission lines and comparing line ratios. Thus little has been done to examine fully the usefulness of these lines for obtaining quiet Sun densities.

4.4.6 Nitrogen Isoelectronic Sequence

At quiet-Sun densities lines from the ions Si VIII and S X are density sensitive. Feldman et al. (1978b) have examined the density-sensitive lines of Si VIII and S X that are available in the wavelength range of the NRL slit spectrograph on *Skylab*. The only useful diagnostic in this wavelength range is the ratio of the 1213.00 and 1196.26 Å lines of S X. These lines are extremely weak and have only been measured in long-exposure spectra taken well above the limb. At a height of 12 arc sec above the limb, this ratio yields an electron density from a single measurement of 1.5×10^9 cm^{-3} at a temperature of 1.3×10^6 K. At 20 arc sec the average of two measurements is 8.2×10^8 cm^{-3}. Because of the weakness of the lines and the need to make observations high above the limb, this ratio is of limited usefulness.

4.4.7 Line Ratios Using Different Ions

Ratios of lines from a single ion produce density determinations with the fewest number of assumptions. Often, however, a suitable ratio is not available. With an observation of a single intercombination line and one or more allowed lines formed at the same temperature, it is often possible to obtain the electron density using a technique first developed by Feldman et al. (1977).

Consider a simple two-level atom with a metastable upper level. The statistical equilibrium equations describing the level populations are then

$$n_1 n_e C_{12} = n_2 (A_{21} + n_e C_{21}), \tag{4.18}$$

and

$$n_1 + n_2 = n_{ion}. \tag{4.19}$$

Solving these two equations for the population of the excited state, we have

$$\frac{n_2}{n_e n_{ion}} = \frac{C_{12}}{A_{21} + n_e C_{12}\left[1 + (\omega_1/\omega_2)\exp(\Delta E_{12}/kT)\right]}, \tag{4.20}$$

where we have used the detailed balance expression relating the collisional excitation and deexcitation rates. For low electron densities, $A_{21} \gg n_e C_{12}$, and we have

$$\frac{n_2}{n_e n_{ion}} \approx \frac{C_{12}}{A_{21}}, \tag{4.21}$$

which is proportional to temperature only, because of the temperature dependence in the collision rates. At high densities $n_e C_{12} \gg A_{21}$, and we have

$$\frac{n_2}{n_e n_{ion}} \approx \frac{1}{n_e \left[1 + (\omega_1/\omega_2)\exp(\Delta E_{12}/kT)\right]}, \tag{4.22}$$

which is inversely proportional to the density. If we assume ionization equilibrium, then we know from equation (2.4) that

$$n_{ion} = \frac{n_{ion}}{n_{el}} A_{el} \, 0.8 \, n_e. \tag{4.23}$$

The number of photons emitted per second in this line from a unit volume is just $A_{21} n_2$. Thus the ratio of the flux in this line from a given volume to the emission measure of the same volume will be proportional to $1/n_e$.

Figure 4.9 summarizes the density sensitivity of the metastable levels in several ions formed in the transition region. At quiet-Sun densities, the C III 1909 Å emission line exhibits density sensitivity. Emission lines from metastable levels in N III, N IV, and O III also begin to show some density sensitivity at quiet-Sun densities.

Application of these diagnostics is more complex than for cases involving the same ion. For the C III 1909 Å line, the best nearby allowed emission line is the Si IV 1403 Å line. Using a simple two-level model for the line, we have

$$F(1403 \text{ Å}) = \frac{1}{4\pi R^2} \frac{hc}{\lambda} 0.8 A_{Si} \left\langle \frac{n_{Si\,IV}}{n_{Si}} C_{lu} \right\rangle_{\log T = 4.85} \int n_e^2 \, dV, \tag{4.24}$$

where the expression in angled brackets is essentially the $G(T)$ function discussed earlier. The flux from the C III line can be similarly written to give

Density Diagnostics

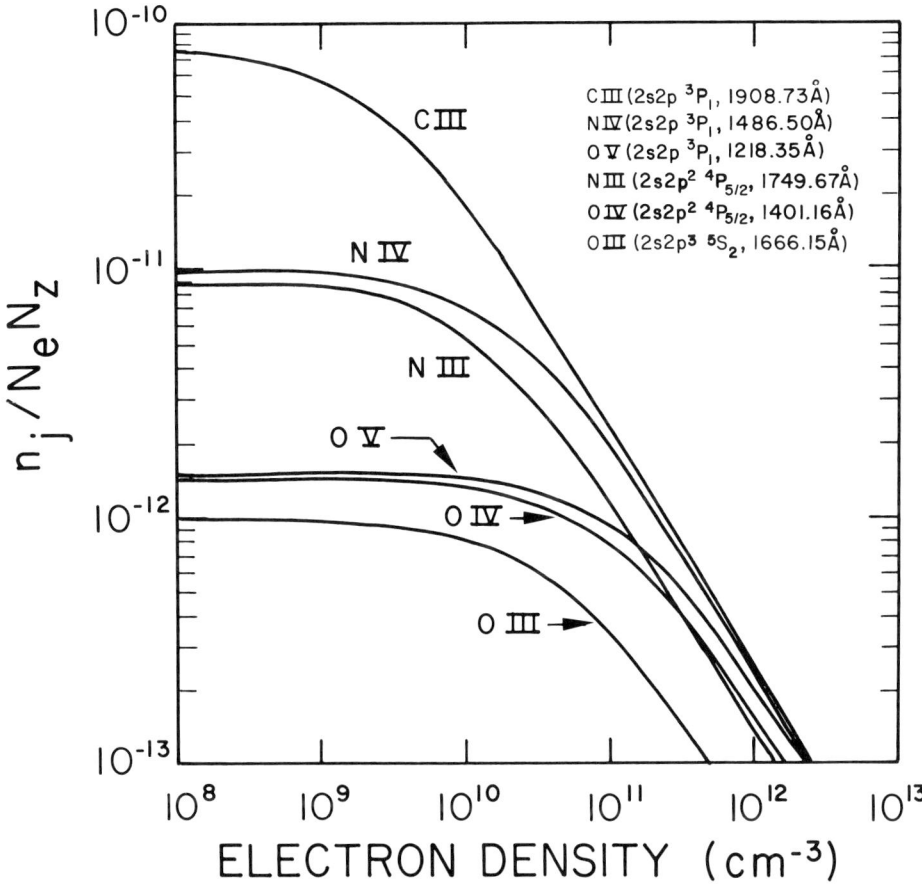

Fig. 4.9. Ratio of the excited level population to the product of the electron density and the ion density for several excited levels that exhibit density sensitivity (after Doschek, 1985).

$$F(1909 \text{ Å}) = \frac{1}{4\pi R^2} \frac{hc}{\lambda} 0.8 A_C \left\langle \frac{n_{\text{C III}}}{n_C} \frac{n_u}{n_{\text{ion}}} \frac{A_{ul}}{n_e} \right\rangle_{\log T = 4.80}$$
$$\times \int n_e^2 \, dV. \tag{4.25}$$

In the low-density limit, we can replace the product $n_u A_{ul}$ with the excitation rate $n_l n_e C_{lu}$, and the expression in the angled brackets is again just the $G(T)$ function. Even at higher densities, it is a sharply peaked function of temperature and thus we can remove it from the integral in the same manner as the $G(T)$ function.

Taking the ratio of these two lines we have

$$\frac{F(1403\,\text{Å})}{F(1909\,\text{Å})} = \frac{1909\,A_{\text{Si}}}{1403\,A_{\text{C}}} \frac{\left\langle \dfrac{n_{\text{Si\,IV}}}{n_{\text{Si}}} C_{lu} \right\rangle_{\log T=4.85}}{\left\langle \dfrac{n_{\text{C\,III}}}{n_{\text{C}}} \dfrac{n_u}{n_{\text{C\,III}}} \dfrac{A_{ul}}{n_e} \right\rangle_{\log T=4.80}} \frac{\int n_e^2\,dV}{\int n_e^2\,dV}.$$

(4.26)

Equation (4.26) highlights the additional assumptions associated with this technique. First, we must know the elemental abundance ratio. Second, we must know the relative ionic abundance as a function of temperature for the two ions. Finally, since the two lines do not form at the same temperature, we must know the shape of the emission measure distribution to divide out the emission measure ratio. These complications are in addition to the need to have an accurate atomic physics model that will correctly predict the population of the metastable level. This calculation is difficult alone, due to the presence of resonances at low energies in the collision rates for the metastable level.

Cook and Nicolas (1979) examined the Si IV 1403 Å to C III 1909 Å line ratio in several sets of quiet solar limb data with positions ranging from 12 arc sec inside to 8 arc sec outside the white-light limb. The electron densities measured with this ratio ranged from 9.3×10^9 to 4.8×10^{10} cm^{-3}, with an average value of 2.1×10^{10} cm^{-3} at a temperature of 63,000 K. These values agree well with densities they determined using lines of C III alone.

A second useful ratio can be formed by comparing the 1892 Å Si III intersystem line with the 1909 Å C III intersystem line. At quiet-Sun densities, the Si III 1892 Å line behaves like an allowed line, so the ratio functions in the same manner as the $I(1403\,\text{Å})/I(1909\,\text{Å})$ diagnostic. One advantage of this ratio is that the two lines are formed at nearly the same wavelength, which minimizes instrumental calibration problems. One disadvantage is that both lines are intersystem transitions. Calculations of their collision rates are subject to the influence of resonances at low energies. Doschek et al. (1978b) first examined the density sensitivity of this ratio. They noted that its sensitivity extends to quite low values of the electron density, making it an excellent candidate for obtaining densities in late-type giant stars.

Cook and Nicolas (1979) obtained quiet-Sun densities for this diagnostic for positions varying from 12 arc sec inside the white-light limb to 12 arc sec outside the limb. The derived densities varied from 6.7×10^9 to 3.6×10^{10} cm^{-3}, with an average value of 1.8×10^{10} cm^{-3} at 63,000 K. This average is in excellent agreement with the value they found using the $I(1403 \text{ Å})/I(1909 \text{ Å})$ ratio. They did note, however, that above the limb the ratio $I(1403 \text{ Å})/I(1909 \text{ Å})$ gave a consistently higher density than $I(1892 \text{ Å})/I(1909 \text{ Å})$ and suggested that this might be due to a coronal contribution to the Si IV line intensity. Si IV is Na-like and thus has a high-temperature tail on its ionization equilibrium distribution. At large heights above the limb, this may contribute appreciable emission to the total line intensity.

For quiet-Sun densities, diagnostically useful line ratios can also be formed using the O III intersystem line at 1666.15 Å. Doschek et al. (1978a) have investigated ratios involving this line and both the Si IV 1403 Å line and the C III 1909 Å line. Their analysis of these lines is similar in spirit, but differs in detail from that of Cook and Nicolas (1979). Rather than computing a normalization factor, as outlined in Section 4.1, for the average value of the temperature-dependent factors as Cook and Nicolas did, they used 0.7 of the peak value and then adjusted for the difference in the widths of the temperature-dependent factors.

The solid curves in Figure 4.10 show the three density-sensitive ratios formed in this manner. As they are plotted, these ratios do not yield consistent densities for the same solar region. Doschek et al. (1978a) argued that this was due to errors in abundance determinations and instrumental calibration. To remove these errors, they noted that at high densities the ratio $I(1909 \text{ Å})/I(1666 \text{ Å})$ approaches a high-density limit. For densities greater than 10^{11} cm^{-3}, the theoretical ratio is near 1. *Skylab* observations of active regions gave a value for this ratio of 2.6. Thus Doschek et al. multiplied each value of the theoretical ratio by 2.7 to produce the correct high-density limit. This produces the dashed curve shown in the figure. Using coronal hole observations, they performed a similar normalization on the ratio $I(1666 \text{ Å})/I(1403 \text{ Å})$. Once this ratio was normalized, active region densities determined using it were used to normalize the C III to Si IV ratio $I(1909 \text{ Å})/I(1403 \text{ Å})$, which is also shown in the figure. This normalization process produced a theoretical curve that is similar to the curve derived by Cook and Nicolas (1979).

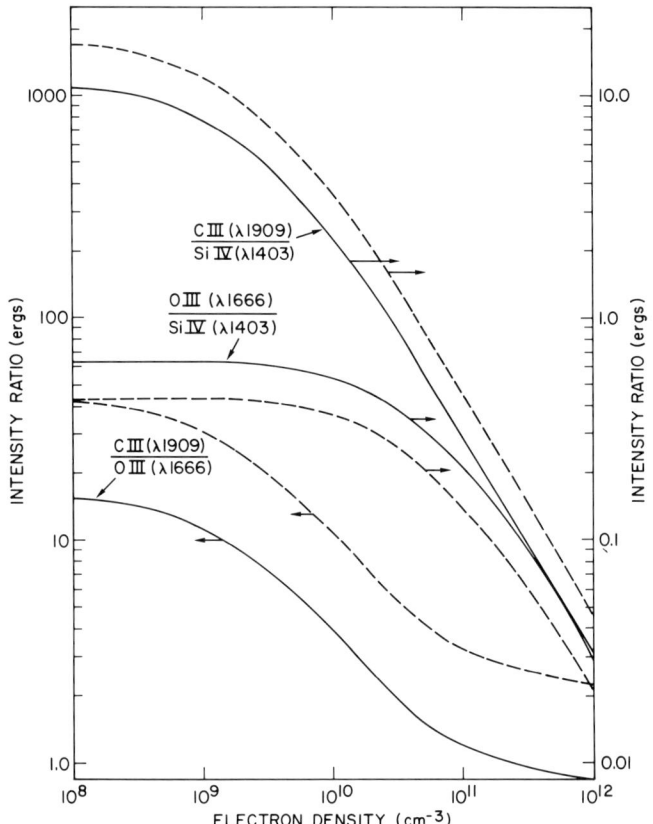

Fig. 4.10. Density-sensitive interspecies line ratios. The arrows refer to the ordinate scale for each curve. Dashed curves have been normalized using the technique described in the text (from Doschek et al. 1978a).

Using the C III 1909 Å to O III 1666 Å ratio, Doschek et al. (1978a) found an average value of 1.9×10^{10} cm^{-3} at a temperature of about 60,000 K for the density between 2 and 4 arc sec above the limb. For the C III 1909 Å to Si IV 1403 Å ratio, they found an average quiet-Sun density at 60,000 K of 1.6×10^{10} cm^{-3}. This determination is based only on observations made at 2 and 4 arc sec inside the white-light limb and above 4 arc sec outside the limb, which avoids any possible opacity effects in the Si IV line.

Note that the shift in the C III 1909 Å to O III 1666 Å ratio is substantial, about a factor of 5 in density along the linear portion of the curve. Some of this shift may be due to differences in the emission measure at the temperature of formation of the two lines.

While the C III 1909 Å to Si IV 1403 Å ratio is clearly an excellent diagnostic of electron densities, there is undoubtedly more uncertainty associated with its use than with ratios from the same ion. The two emission lines originate from volumes of plasma that only partially overlap in temperature, leading to the possibility that significant portions of one line could originate in a completely different structure along the line of sight than the other line. Abundance differences could also affect the ratio. In addition, the normalization process used by Doschek et al. (1978a) must be used with some caution. To the extent that errors in the instrumental calibration are removed, it is instrument specific. Since it also removes differences in emission measure between the two lines, the normalization process implicitly assumes that the slope of the emission measure in an active region seen at the limb is the same as that of a quiet region seen at the limb. Given the well-known complex structure of active regions at the limb as shown for example in the NRL *Skylab* spectroheliograph images (Feldman et al., 1987), there is a significant chance of introducing errors in this process. Thus, using carefully constructed unnormalized theoretical line ratios in the manner described by Cook and Nicolas (1979) is probably a better approach. It also makes the theoretical curves useful for any low-density collisionally excited plasma, such as the outer atmospheres of late-type stars.

The availability of UV spectra from late-type stars has stimulated a careful reexamination of all these line ratios. Improved theoretical ratios have been published for all the useful solar interspecies electron density diagnostics and a number of additional ratios (Keenan et al., 1987, 1988, 1990). Many of these ratios are useful in stars of later spectral type than the Sun, because the continuum intensity decreases relative to the emission lines.

4.4.8 *Transition Region Density*

Figure 4.11 summarizes the density determinations we have discussed in this section. Here we have plotted the electron density as a function of the temperature at which it was measured. Most of the determinations are from limb data. Also plotted are constant pressure lines with values of n_eT of 10^{14}, 10^{15}, and 10^{16} cm^{-3} K.

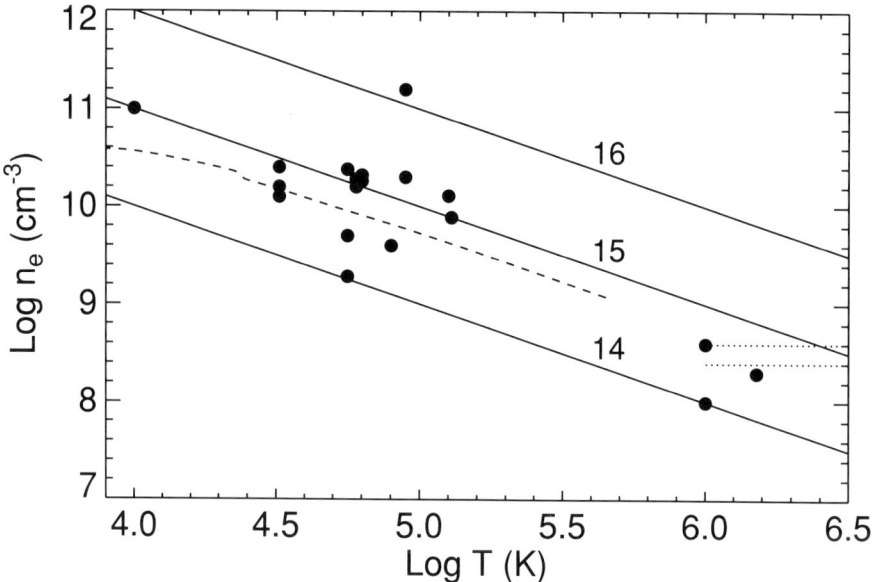

Fig. 4.11. Quiet-Sun electron densities as a function of temperature. The solid lines show densities for a constant pressure atmosphere with the values of $\log n_e T$ indicated. Dotted lines show approximate solar maximum and solar minimum coronal values, and the dashed line shows the electron density distribution in the Vernazza *et al.* (1981) average model.

The dashed curve in the plot shows the electron density tabulated for the average quiet Sun model of Vernazza *et al.* (1981). This model should be considered highly uncertain above temperatures of 40,000 K ($\log T = 4.6$), where the Lyman α line has become optically thin. At high temperatures, the two horizontal lines are the coronal electron densities at a height of 0.01 $R_\odot$ above the limb at solar maximum and minimum tabulated by Allen (1973). This height is about 10 arc sec above the limb, and thus the measurements should be roughly comparable to the upper transition region determinations.

While there is considerable scatter in the measurements, the trend in the density determinations suggests that the transition region is at constant pressure. Moreover, the lack of any large discontinuity between the independently determined densities in the Vernazza *et al.* (1981) model and the transition region densities suggests that much, if not all, of the emitting plasma at transition region temperatures physically connects to the lower atmospheric layers. That is, there is no evidence in the density determinations for magnetically isolated transition region structures. At high temperatures, the densities are in good agreement with the coronal determinations.

The scatter in the measurements at temperatures below 10^5 K is due to errors in atomic data, the large uncertainties produced by using diagnostics near their low-density limit, and actual variations in density from region to region on the Sun. For example, the value of 1.58×10^{11} cm^{-3} at 90,000 K ($\log T = 4.95$) is from the ratio $I(1423.9$ Å$)/I(1416.9$ Å$)$ of S IV. At these densities the ratio is near the low-density limit and the 1417 Å line is weak, leading to a highly uncertain determination.

In examining diagnostics involving the C III ion, Cook and Nicolas (1979) tabulated individual density determinations of three separate quiet solar limb data sets. Within each set of data the average densities determined from each ratio agreed to better than 50%. For many individual ratios, the densities determined at different heights in the same quiet region differed by up to a factor of 2. The average densities for each region differed by up to a factor of 2 from the averages for other regions. Thus it appears that the better diagnostics can determine densities to better than a factor of 2, while the actual variations in the quiet-Sun electron density at a given temperature in the transition region are of the order of a factor of 2.

Although the trend over the entire temperature range covered by the figure supports a constant pressure transition region, many of the best determinations at temperatures below 10^5 K seem to cluster near a value of 2×10^{10} cm^{-3}. Determining whether this trend continues at higher temperatures will require the application of density diagnostics in the 10^5–10^6 K range. These determinations must await new high-resolution data in the EUV wavelength range.

When combined with the emission measures discussed in Section 4.2, electron density determinations provide us with estimates of characteristic size scales, which we can then compare with the sizes of emitting regions seen in high spatial resolution images. At 60,000 K, the average electron density is about 1.8×10^{10} cm^{-3}. The quiet network emission measure shown in Figure 4.1 at the same temperature is about 6.9×10^{25} cm^{-5}. Dividing this by the square of the electron density, we have a characteristic length of only 2.1 km. If we assume that the transition region is at constant pressure, then at 2×10^5 K the characteristic length from the network emission measure in Figure 4.1 is 7 km. At 4×10^5 K it is 74 km, and at 8×10^5 K it is 760 km. Only this last value, which is a little more than 1 arc sec, is resolvable with current instrumentation.

Clearly, the actual structure of the transition region is quite complex. The limb observations discussed in Section 3.2 show that at 10^5 K, significant emission extends over more than 2 arc sec, 1450 km. To yield emission over these lengths at the limb, there must either be great numbers of small emitting centers at different heights or the emitting plasma must be in the form of thread-like structures.

This structuring immediately raises questions about the validity of electron density determinations in multicomponent plasmas. Doschek (1984) investigated density diagnostics for an isothermal plasma with one high-density region and one low-density region. He concluded that the presence of a small volume of high-density plasma would result in a measured electron density that lies between the densities of the two regions. For density diagnostics involving the ratio of an intersystem line to an allowed line, the density would be close to that of the high-density region, because the allowed line would be formed primarily in the high-density region. For ratios involving two intersystem lines, the derived density would be smaller. Almleaky *et al.* (1989) have examined the more general problem of an isothermal inhomogeneous plasma.

4.5 Temperature Diagnostics

Because the transition region is highly structured, any line of sight observed in the UV or EUV will contain plasma over a considerable range of temperatures. On the disk this range will include the atmosphere from the corona to the chromosphere. Above the limb some of the lower atmosphere will be eliminated from the line of sight. We see individual structures in different UV and EUV emission lines because each ion is present over only a narrow temperature range. If the plasma is in ionization equilibrium, that temperature range is narrow enough that a relatively precise temperature can be assigned to the emission from each line.

Observations show, however, that the transition region is highly dynamic. Thus it is likely that departures from ionization equilibrium are common. Measuring the temperature of an emitting region and comparing it with the temperature expected in ionization equilibrium is one way to refine further our understanding of the structure and dynamics of the transition region.

In principle measuring the temperature is straightforward. One simply measures the ratios of a pair of lines from the same ion that have different Boltzmann factors, $\exp(-\Delta E/kT)$. Consider, for example, the resonance and intersystem lines of O V. At 2.2×10^5 K, the peak temperature of formation of the ion in equilibrium, the ratio of the Boltzmann factors for the 629 Å resonance line to the 1218 Å intersystem line is 0.60. At twice that temperature it is 0.78, and at half that temperature it is 0.35. For lower temperatures the ratio is even more sensitive. Because of the larger difference in excitation energies, the ratio of the Boltzmann factors for the 172 Å O V line to the 1218 Å line is even more sensitive to temperature.

In practice these and other line pairs like them are virtually impossible to apply to solar observations. The problem is the large difference in wavelength between the two lines. No single spectrograph can accurately record both lines. Combining observations from two instruments is difficult because the calibrations are likely to be different. Moreover, the two instruments never observe exactly the same area on the Sun.

In some atomic systems, however, there are temperature-sensitive line ratios in which the two lines are close in wavelength. The Na-like ions offer the best examples of this technique (Flower and Nussbaumer, 1975a). For these ions the relative intensities of the $3s$–$3p$ and $3p$–$3d$ levels are temperature sensitive. Both the $3p$ and the $3d$ levels are excited from the $3s$ levels, but the $3s$–$3d$ excitation energy is about twice the $3s$–$3p$ energy. Because the $3d$ levels then decay to the $3p$ levels, however, the emission lines produced as a result of the excitation are close in wavelength to those produced by the $3s$–$3p$ excitations.

Flower and Nussbaumer (1975a) first applied this technique to the Na-like ions Si IV and S VI. They calculated radiative and collisional excitation rates using the distorted wave approximation and computed the temperature-sensitive line ratios. Comparisons with low spectral resolution solar data, however, showed large discrepancies between the theoretical and observed ratios, probably due to blends.

Keenan et al. (1986b) used more accurate close-coupling calculations by Dufton and Kingston (1987) to examine temperature-sensitive ratios in Al III and Si IV. For Al III there are four lines in the solar spectrum that are useful for temperature determinations. The $3d\ ^2D_{3/2,5/2} \rightarrow 3p\ ^2P_{3/2}$ line at 1611.9 Å and the $4s\ ^2S_{1/2} \rightarrow 3p\ ^2P_{1/2}$ line at 1379.7 Å are the result of excitations from the $3s$ level and thus represent the excitation of a photon with an energy of about 860 Å. A weaker line at 1605.76 Å due to the decay to the $^2P_{1/2}$ level has also been observed, but is blended with a lower temperature line (Doschek and Feldman, 1987). These lines can be compared with the resonance lines produced by the $3p\ ^2P_{1/2} \rightarrow 3s\ ^2S_{1/2}$ and the $3p\ ^2P_{3/2} \rightarrow 3s\ ^2S_{1/2}$ transitions at 1862.8 and 1854.7 Å.

From these lines, four temperature-sensitive ratios can be formed:

$$R_1 = I(1611.9\ \text{Å})/I(1862.8\ \text{Å}), \tag{4.27}$$

$$R_2 = I(1611.9\ \text{Å})/I(1854.7\ \text{Å}), \tag{4.28}$$

$$R_3 = I(1379.7\ \text{Å})/I(1862.8\ \text{Å}), \tag{4.29}$$

and

$$R_4 = I(1379.7\ \text{Å})/I(1854.7\ \text{Å}). \tag{4.30}$$

Keenan et al. (1986b) examined these ratios in the available NRL Skylab spectrograph data. Their average values for R_1 and R_2 in the quiet Sun were 2.0×10^{-2} and 7.5×10^{-3}, respectively, leading to an average value of $\log T$ of 4.4 for both ratios. For R_3 and R_4 they found average quiet-Sun values of 6.0×10^{-3} and 4.5×10^{-3}, leading to average values of $\log T$ of 4.6 and 4.7. In equilibrium the Al III abundance peaks at a $\log T$ of 4.6. For a coronal hole, a surge, and an active region, they found values as high as 5.0 and 5.1 based on R_3 and R_4. These higher values are probably the result of blending of the 1379.7 Å line with the 1379.61 Å Fe II line. Thus the R_1 and R_2 ratios are the best for solar spectra.

Doschek and Feldman (1987) also examined the R_2 ratio in the NRL Skylab data. They performed a more careful determination of the line intensities, even accounting for the opacity in the 1854.7 Å resonance line and removing the contribution from the $^2D_{3/2} \rightarrow {}^2P_{3/2}$ transition. The resulting quiet-Sun ratio was 6.1×10^{-3} using the Doschek et al. (1976b) instrumental calibration and 1.1×10^{-2} using the Nicolas et al. (1977) calibration. Combining these measurements and the theoretical curves of Keenan et al. (1986b) gives values for $\log T$ of about 4.3 and 4.5, respectively.

Thus all the determinations yield values for the temperature that are below the equilibrium temperature. The sizes of the departures are, however, relatively small, and could be accounted for by calibration errors. On the other hand, physical effects such as heating or cooling of the plasma and flows could alter the temperature of formation of the ion and account for the observed discrepancy. We will consider that topic in greater detail in a later chapter.

Keenan et al. (1986b) also calculated the temperature sensitivity of the R_1 and R_2 ratios for Si IV. The emission lines involved are at 1128.3, 1393.8, and 815.0 Å. Only the 1393.8 Å line is within the wavelength range of the NRL *Skylab* spectrograph. The other two lines fall within the wavelength range of the lower spectral resolution Harvard spectroheliometer flown on *Skylab*. Because of the different fields of view and the lack of a detailed intercalibration between the two instruments, it is not wise to form a ratio with lines from each of them. Dupree and Reeves (1971) have reported quiet-Sun measurements from the *OSO-IV* satellite for the lines in the ratio $R_2 = I(1128.3 \text{ Å})/I(1393.8 \text{ Å})$. Their observations give a value for this ratio of 0.17, which is above the high-temperature limit for the theoretical ratio and well above the value of 0.028 expected at $\log T = 4.8$, the equilibrium temperature of formation of Si IV. This large discrepancy is probably due to blending with other lines due to the roughly 2.0 Å spectral resolution of the instrument. Application of this ratio must therefore await future observations.

Doschek and Feldman (1987) have suggested that the S III lines at 1328.12, 1328.52, and 1343.539 Å may be useful temperature diagnostics. These lines are produced by transitions from the $3s^23p4p$ to the $3s3p^3$ levels. On the Sun the upper levels are probably excited from the $3s^23p^2$ ground state, and thus represent a photon with a wavelength of about 620 Å. These lines could be compared with the strong S III line at 1200.972 Å. Application of this potential diagnostic must await detailed atomic physics calculations for the S III system.

Lines of the Na-like ion Mg II are also temperature sensitive. Feldman and Doschek (1977a) examined NRL *Skylab* observations of the ratio of the two $3d \to 3p$ lines at 2790.768 and 2797.989 Å to the two $3p \to 3s$ resonance lines at 2802.698 and 2795.523 Å at heights ranging from 0 to 20 arc sec above the limb in a quiet region and an active region. For the quiet region, only the data at 4, 6, and 8 arc sec are free of blends, unmasked by continuum emission, and strong enough for accurate measurements. Opacity is still, however, a problem. In addition, the theoretical ratios must include radiative excitation from the intense chromospheric radiation field. This introduces a pressure dependence into the curves. Using only the quiet region data at 8 arc sec, which should show the least opacity, Feldman and Doschek estimated a temperature of 14,000 K at a pressure of 2×10^{15} cm^{-3} K and a temperature of 18,000 K at a pressure of 6×10^{14} cm^{-3} K. The density diagnostics discussed in the last section suggest that the pressure at 60,000 K is 1.1×10^{15} cm^{-3} K, indicating that the smaller temperature is preferable. Both are quite close to the value of 16,000 K for the peak equilibrium abundance of Mg II determined by Shull and Van Steenberg (1982a) and the value of 12,000 K determined by Arnaud and Rothenflug (1985).

Keenan and Aggarwal (1989) examined temperature-sensitive line ratios in the C I sequence ion O III. They found that the ratios $I(703 \text{ Å})/I(599.6 \text{ Å})$ and $I(834 \text{ Å})/I(599.6 \text{ Å})$ yield useful temperature diagnostics. Using these ratios and the quiet-Sun disk and limb data published by Vernazza and Reeves (1978), they obtained temperatures ranging from $\log T = 4.72$–5.19. The O III fractional abundance peaks at $\log T = 5.0$ (Arnaud and Rothenflug, 1985). Thus the measured values scatter around the equilibrium value.

None of the available temperature diagnostics show any evidence for significant departures of the temperature of formation of UV and EUV emission lines from the value expected in ionization equilibrium. This is perhaps not surprising when we consider the amount of fine structure being integrated over in a typical observation. If the observed regions are in ionization equilibrium, then the temperature-sensitive line ratio technique will be heavily biased toward the equilibrium value. Feldman et al. (1978a) have shown that for this case there must be an overwhelmingly larger amount of emitting material at temperatures significantly larger or smaller than the equilibrium temperature to result in even a small change in the ratio. The reason is simply the sharply peaked shape of the $G(T)$ function for each emission line. Except possibly for observations above the limb, it is difficult to isolate a particular structure from foreground and background emission. Thus, if the plasma in most solar locations is in ionization equilibrium, temperature-sensitive line ratios would be strongly biased toward their equilibrium values.

On the other hand, there is strong observational evidence for time-dependent fluctuations in emission line intensities. Furthermore, as we will show in the next chapter, there is strong observational evidence for both steady and impulsive mass motions in the transition region. These fluctuations should lead to departures from ionization equilibrium. They may, however, be in small enough regions or short-lived enough to escape detection in the time averaged observations we have discussed. Some of the spectra we have discussed required exposure times of up to 640 s, possibly leading to averaging over short-lived departures from equilibrium.

5

High-Resolution Spectroscopy

High spectral resolution observations reveal the details of emission-line profiles. These line profiles can, in principle, tell us a great deal about the physical conditions in the line forming region. This chapter summarizes the current data and discusses what that data reveal about the dynamics of the transition region.

5.1 Emission-Line Profiles

In Chapter 2 we noted that for an optically-thin emission line the volume emissivity of the plasma is

$$\varepsilon_\nu = h\nu_{ji} A_{ji} n_j \psi_\nu. \tag{5.1}$$

Many processes can contribute to the emission profile ψ_ν. For the optically-thin UV emission lines formed in the solar transition region and corona, Doppler broadening is the most important process. Because each atom is in thermal motion, the observed frequency of any photon it emits at a fixed frequency in its own frame of reference will be different. Each atom has its own velocity, and hence Doppler shift, so the net effect is to spread the line emission in wavelength. The same number of photons is still emitted, however, so the total line strength remains the same.

Consider an atom with velocity component v_l along the line of sight. To lowest order in v/c, the change in frequency associated with this velocity is given by

$$\nu - \nu_0 = \frac{\nu_0 v_l}{c}, \tag{5.2}$$

where ν_0 is the rest-frame frequency. To arrive at the line profile, we integrate over the number of atoms in each velocity interval.

The number of atoms in the velocity range v_l to $v_l + dv_l$ is given by the Maxwellian distribution

$$dn(v_l) = n \left(\frac{M}{2\pi kT}\right)^{1/2} \exp\left(-\frac{Mv_l^2}{2kT}\right) dv_l, \qquad (5.3)$$

where M is the mass of the atom. Setting

$$v_l = \frac{c(\nu - \nu_0)}{\nu_0} \qquad (5.4)$$

and

$$dv_l = \frac{c\, d\nu}{\nu_0}, \qquad (5.5)$$

we can integrate equation (5.3) to obtain the profile function

$$\psi_\nu = \frac{1}{\pi^{1/2} \Delta\nu_D} \exp\left[-\frac{(\nu - \nu_0)^2}{\Delta\nu_D^2}\right]. \qquad (5.6)$$

Here the Doppler width $\Delta\nu_D$ is defined as

$$\Delta\nu_D = \frac{\nu_0}{c}\left(\frac{2kT}{M}\right)^{1/2}. \qquad (5.7)$$

If the profile function is written in terms of wavelength, the Doppler width becomes

$$\Delta\lambda_D = \frac{\lambda_0}{c}\left(\frac{2kT}{M}\right)^{1/2}. \qquad (5.8)$$

In these expressions the temperature is strictly speaking an ion temperature, since it is the ions that are in thermal motion. At the temperatures and densities of the transition region, however, the electrons and ions rapidly equilibrate. For electron-ion collisions and solar conditions, the thermal equilibration rate constant $\nu^{e/i}$ (s^{-1}) is given by (Book, 1983; Trubnikov, 1965)

$$\nu^{e/i} = \frac{4 \times 10^{-3} Z^2 \lambda n_i}{\mu T^{3/2}}, \qquad (5.9)$$

where Z is the mean ionic charge, μ is the mean ionic mass in proton masses, λ is the Coulomb logarithm, and the temperature is in Kelvins. For a solar transition region plasma, $\lambda \approx 19$, $\mu \approx 1.2$, $Z \approx 1.1$, and $n_i \approx n_e$. Thus the characteristic equilibration time is given by

$$1/\nu^{e/i} \approx 13\, T^{3/2}/n_e. \qquad (5.10)$$

For a typical electron density of 10^{10} cm^{-3} and temperature of 10^5 K, the equilibration time is about 0.04 s. Assuming a single temperature for the quiet transition region plasma is therefore reasonable.

Along with thermal motions, there also may be motion associated with large- and small-scale velocity fields. When the scale of these motions is small compared with the instrumental resolution, they are often accounted for by adding a component to the Doppler width

$$\Delta\lambda_D = \frac{\lambda_0}{c}\left(\frac{2kT}{M} + \xi^2\right)^{1/2}, \tag{5.11}$$

where ξ is the most probable nonthermal velocity. This assumes that the nonthermal velocities also have a Gaussian distribution. Occasionally the nonthermal velocity will be expressed in terms of a root-mean-square velocity

$$v_{\rm rms} = (3/2)^{1/2}\xi. \tag{5.12}$$

There are of course other mechanisms that broaden spectral lines, such as natural broadening and collisional broadening. When natural broadening, which produces a Lorentz profile, is combined with Doppler broadening, the result is a more complex form for the line profile described by a Voigt function (e.g., Mihalas, 1978). Within about five Doppler widths of line center, however, a Gaussian function represents the profile well. For optically thin UV emission lines, there is little need to consider the more complicated form for the profile.

5.2 Line Width Observations

At low to moderate spatial resolution, all emission lines formed in the quiet solar transition region are broadened in excess of their thermal width (Brueckner and Moe, 1972; Boland et al., 1975; Doschek et al., 1976b). Moreover, with almost no exceptions, the line profiles are Gaussian in shape. Figure 5.1 shows a series of typical optically-thin line profiles observed near the limb with the NRL spectrograph on *Skylab*. These profiles are for the 1486.5 Å N IV emission line, which forms at a temperature of about 1.3×10^5 K. Although there are fluctuations due to film grain, Gaussian profiles provide an excellent fit to these lines.

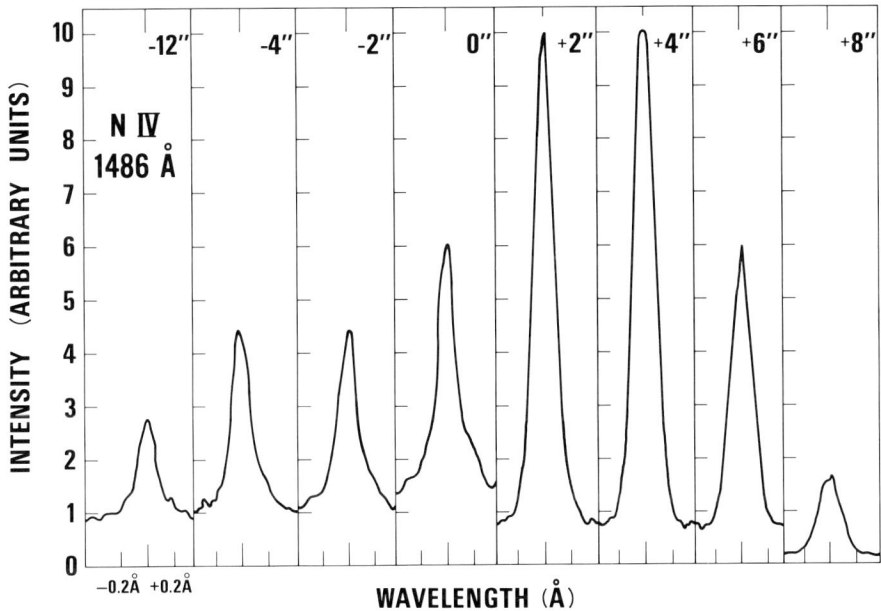

Fig. 5.1. Profiles of the N IV 1486 Å emission line at different positions relative to the solar white-light limb (from Doschek et al., 1976b).

5.2.1 Variation with Temperature

Line profile observations such as those shown in the figure have been made in many lines spanning the temperature range from the chromosphere through the corona. Since the profiles are Gaussian in shape, a measurement of the full width at half-maximum W of each line coupled with the temperature of formation of the ion immediately yields an estimate of the nonthermal velocity through the relation

$$W = \left[4 \ln 2 \left(\frac{\lambda}{c} \right)^2 \left(\frac{2kT}{M} + \xi^2 \right) \right]^{1/2}. \qquad (5.13)$$

Of course the additional broadening introduced by the instrument must first be removed before this measurement can be made. Often it is included as an instrumental velocity in the expression in parentheses.

Figure 5.2 shows the behavior of ξ as a function of the temperature of formation of the emission line. We have only included optically-thin emission lines in the plot. Thus we have excluded

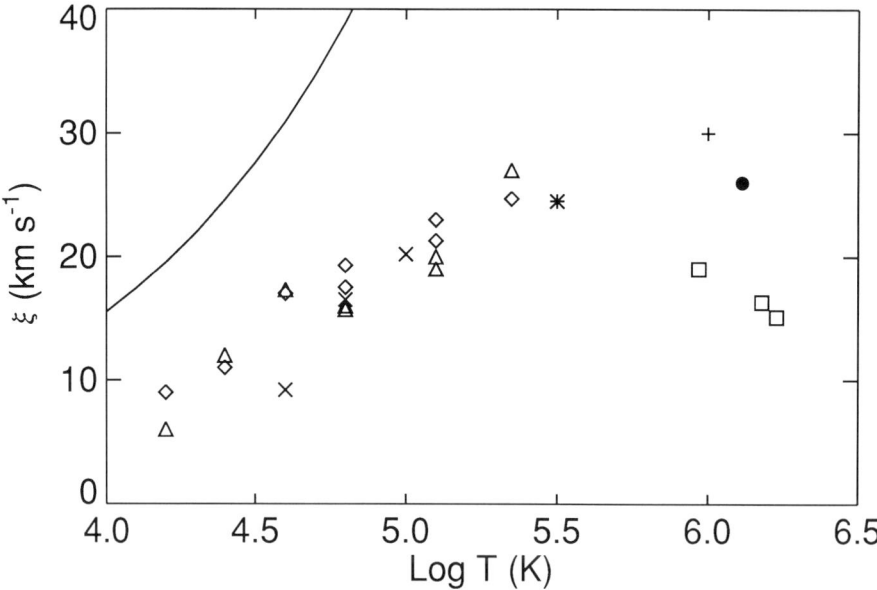

Fig. 5.2. Nonthermal broadening of optically-thin emission lines as a function of temperature. The data represented with triangles are from Doschek et al. (1976b), with diamonds are from Mariska et al. (1978), with crosses are from Boland et al. (1975), with a star is from Bonnet et al. (1978), with a plus is from Feldman and Behring (1974), with the squares are from Cheng et al. (1979), and with a filled circle is from Hassler et al. (1990).

some strong transition region lines, such as the resonance lines of C IV, since they are clearly broadened by opacity near the limb. Also plotted on the figure is the local sound speed, calculated assuming a ratio of specific heats of 5/3. In the transition region the figure shows that there is a strong correlation between the nonthermal velocity and the temperature of line formation. The nonthermal velocity is, however, subsonic everywhere.

Data for the corona are very limited. The point on the plot from the work of Feldman and Behring (1974) is from a spectrograph that observed the entire solar disk. Thus there may be a contribution from active regions. In addition, the authors only quote an average value for several lines that span a range in temperature, some of which may have an opacity contribution to the broadening. The three coronal data points from Cheng et al. (1979) are for optically-thin UV forbidden lines and should be more reliable.

The final coronal data point from Hassler et al. (1990) is for a disk measurement of the Mg X 625 Å line and should be taken as an excellent measurement at that temperature. If we take the Cheng et al. (1979) and Hassler et al. (1990) data as the most reliable at coronal temperatures, then it is clear from the figure that at some temperature in the upper transition region the nonthermal velocity must peak and then either remain roughly constant or begin to decline.

5.2.2 The C IV Line

Until quite recently data on line profiles observed at high spatial resolution, ~ 1 arc sec, were limited. Now, however, considerable data exists on the resonance lines of C IV at 1548 and 1551 Å (Dere et al., 1984, 1987). These lines form at a temperature of about 10^5 K and thus should be excellent diagnostics of transition region dynamics. At the solar limb, however, they suffer from some broadening by opacity. Thus only observations well inside the limb are useful for nonthermal motion studies.

When observed at high spatial resolution, the C IV resonance line profiles are still Gaussian in shape, but exhibit a range of widths. Figure 5.3 shows the distribution of nonthermal velocities in the C IV resonance lines measured by Dere et al. (1987) using 1 arc sec resolution data obtained with the NRL HRTS experiment on *Spacelab 2*. This distribution has an average value for ξ of 16 km s^{-1}. The average value for the set of line width measurements corrected for instrumental broadening was 0.195 Å. This yields a value for ξ of 19 km s^{-1}. It differs from the average value of ξ because the nonthermal velocity is a nonlinear function of the line width.

Earlier low-resolution measurements of the widths of the C IV resonance lines are higher, in the range from 0.2 to 0.25 Å (Boland et al., 1975; Feldman et al., 1976b; Kjeldseth Moe and Nicolas, 1977; Bruner and McWhirter, 1979; Athay and White, 1980). Dere et al. (1987) attribute this difference to three effects. First, their data set excluded the small regions in the quiet Sun with extremely broad non-Gaussian profiles. In lower spatial resolution data these profiles would automatically be included in the average. Second, the 1 arc sec resolution data begins to resolve the transition region flow fields. Additional broadening due to differences in Doppler shift from point to point is thus not included. Finally, the nonlinear

120 *High-Resolution Spectroscopy*

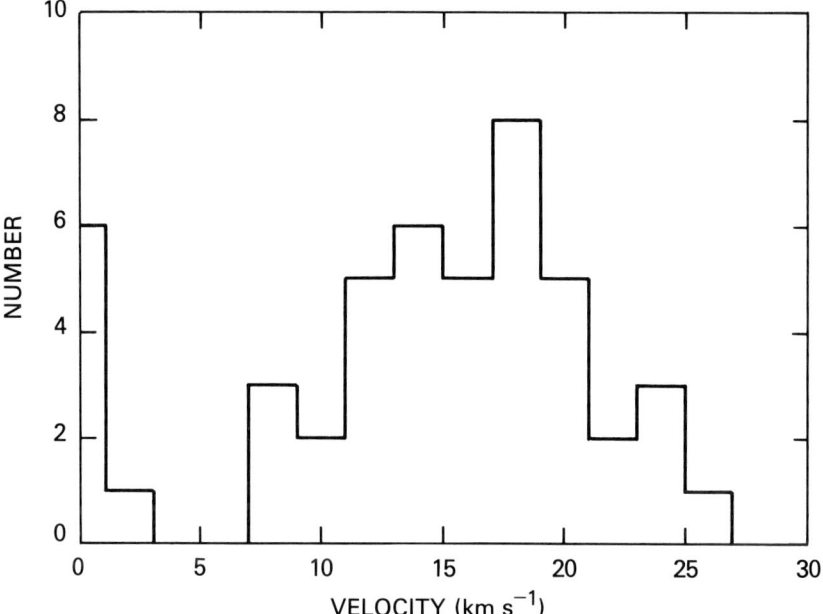

Fig. 5.3. The distribution of nonthermal velocities in the C IV resonance lines from observations with 1 arc sec spatial resolution (from Dere et al., 1987).

dependence of the derived value of ξ on the observed line width means that higher values for ξ will be obtained if the average line width is first computed.

Although the distribution of values for ξ in the C IV resonance lines peaks at 16 km s^{-1}, there are some data points with values at or near zero. These data points raise the possibility that if the instrumental spatial resolution was at the size of the individual structures, the nonthermal broadening might disappear completely. This is clearly not the case at the 1 arc sec resolution of the observations presented in Figure 5.3.

5.2.3 Relation to the Cell-Network Structure

Observations of the relationship between the nonthermal broadening and the brightness of the emitting element are contradictory. Athay et al. (1983) examined quiet-Sun observations obtained in the C IV 1548 Å line with the UVSP on *SMM* at 3 arc sec resolution. They found no relation between the intensity and the width of the emission line. On the other hand, Dere et al. (1984) examined quiet-Sun observations obtained in the C IV 1548 and 1550 Å lines with the NRL HRTS rocket experiment at 1 arc sec resolution and found a strong correlation between the intensity and the width of the line, with bright features having greater widths.

Why these two sets of quiet-Sun measurements should disagree is unclear. Dere et al. (1987) found that the characteristic size of the emitting structures in the C IV resonance line is about 3 arc sec. Thus the *SMM* observations will be partially compromised by averaging over more than one discrete structure. This may be enough to destroy the correlation, but additional data need to be analyzed before a firm conclusion is possible. If indeed the regions with the weakest emission also have the narrowest profiles, then it adds support to the argument that fully resolved structures may show little or no nonthermal broadening (Dere et al., 1984).

5.2.4 Center to Limb Observations

Studies of the center to limb behavior of the widths of emission lines formed in the transition region have been limited. Feldman et al. (1976b) examined one set of network and cell data obtained with the NRL slit spectrograph on *Skylab*. They concluded that the widths of the optically-thin lines of N IV, O IV, Si IV, and C IV were the same in cell interiors and the network and agreed with limb measurements. The only exceptions were the strong lines of C IV at 1548 and 1550 Å and Si IV at 1394 Å, which showed some broadening at the limb due to opacity. *OSO 8* C IV emission line observations showed similar broadening at the extreme limb (Shine et al., 1976).

Roussel-Dupré et al. (1979) performed a more detailed analysis of the *OSO 8* center to limb observations in the 1393 Å line of Si IV. They found that there was an increase in line broadening from the center of the disk to the limb. The increase could be explained by an integrated optical depth over the line of 0.22 ± 0.16, combined with an isotropic microturbulence with a root-mean-square velocity of 13.08 ± 2.08 km s^{-1}. The fit to the data could be improved by assuming that the horizontal component of the velocity was larger than the vertical component, but the differences were small. Roussel-Dupré et al. (1979) were unable to carry out a similar analysis of the C IV 1548 Å line data because of greater scatter in both the limb brightening and line broadening data. Thus the current data appear to indicate that the nonthermal broadening is isotropic.

While on the disk the average width of the optically-thin UV emission lines formed in the transition region shows no systematic variations with position, above the limb there is a systematic trend: the line widths increase with increasing height above the white-light limb. This trend was first noted by Doschek et al. (1977) in emission lines of Si II and C II, which form in the lower transition region. Nicolas et al. (1977) noted the same tendency in optically-thin emission lines of Si III, and Mariska et al. (1978) also found it in lines of C III and O III.

To quantify further this effect, Mariska et al. (1979) examined long-exposure spectra taken above quiet solar regions with the NRL spectrograph on *Skylab*. Their observations covered heights from 8 to 20 arc sec above the limb, and showed that the nonthermal broadening increases with increasing height above the white-light limb for all emission lines formed in the temperature range from $\sim 40,000$ to $\sim 2 \times 10^5$ K. At 12 arc sec above the limb, the nonthermal velocity in the C IV resonance lines is around 33 km s^{-1}, roughly twice the disk value of 16 km s^{-1} determined by Dere et al. (1987). While there is considerable scatter in the observations at 8 and 12 arc sec above the limb, there is a tendency for the nonthermal broadening at any height to increase with temperature of formation in the same manner that it does in the observations at the limb. The data are simply shifted to higher velocities at each temperature. All the velocities, however, remain subsonic.

Although these increases in nonthermal broadening in transition region lines are quite large, it is important to remember that the emitting plasma being measured does not represent the dominant physical conditions above the limb. At 12 arc sec above the limb, the emission in a typical transition region line is reduced by roughly a factor of 50 from its peak value at 2–4 arc sec above the limb (e.g., Mariska et al., 1978). This emission is probably due to extended fine-scale structure such as that seen at the limb in C IV filtergrams (e.g., Dere et al., 1987). If these structures are the extensions of the dominant structural element in the transition region, then the increased broadening may provide additional clues about the dynamics and energetics of the transition region. Inferring the details of coronal heating and dynamics from emission in transition region lines observed well above the limb, however, is hazardous until we understand the structural connections between the two parts of the atmosphere more completely.

Recently, Hassler et al. (1990) obtained line profile measurements as a function of height above the limb in the resonance lines of Mg X at 609 and 625 Å. Their observations extend to more than $0.2\,R_\odot$ above the limb (~ 200 arc sec), and show the same trend in the nonthermal broadening. Between the limb and $0.1\,R_\odot$ above it, the nonthermal broadening increases. Above $0.1\,R_\odot$, the statistics are poorer, but there is some evidence that the broadening may reach a roughly constant value. The fact that the broadening in a coronal line follows the same trends as that in the transition region lines suggests that they may be related by a common mechanism.

5.2.5 Energetic Implications

By far the simplest interpretation of the excess line widths is to assume that they are thermal. A width of 0.20 Å for the C IV resonance lines then implies a temperature of about 4×10^5 K. If the transition region plasma is in equilibrium, the temperature of formation of the C IV resonance lines is 10^5 K. Strictly speaking, the temperature of the peak in the ionization balance is an electron temperature, since electron collisions determine the ionization balance. On the other hand, the temperature determined from the width of the lines is an ion temperature, since it is a direct measurement of the motions of the ions. We showed earlier, however, that the equilibration time for the electrons and ions is only about 0.04 s. Thus maintaining separate electron and ion temperatures would require a steady energy source. Moreover, the power required is quite large. Bruner and McWhirter (1979) estimate that the power required to maintain an electron temperature of 10^5 K against an ion temperature of 5×10^5 K is at least 10^3 times greater than the radiative losses in this region of the atmosphere. Thus such a persistent temperature separation would have profound implications for the overall energy balance of the transition region.

A second possible thermal explanation for the widths of the line profiles is to assume that static ionization equilibrium does not apply and that the electron and ion temperatures are both 4×10^5 K for the C IV lines. This would require rapid heating of the plasma so that the electron temperature was 4×10^5 K, but the ionization balance was still representative of a 10^5 K plasma. It is difficult to envision a mechanism for doing this to a static plasma. Again, the energy requirements are quite large and the process must be relatively fast, since the time for the ionization to catch up is short. On the other hand, a steady flow of plasma through a steep temperature gradient could accomplish this naturally. This should lead to a correlation between the width of the C IV resonance lines and the Doppler shift, which we will discuss in the next section.

Although the exact nature of the nonthermal broadening is uncertain, it is clear that the measured velocities represent considerable energy. If the atmospheric region where an emission line forms has a mass density ρ, and the motions that produce the broadening are isotropic, then the mechanical energy density associated with the motions is given by

$$E = \frac{3}{2}\rho v_{\text{rms}}^2, \tag{5.14}$$

where the factor of 3/2 comes from the assumption that the motions are isotropic. For a quiet-Sun electron pressure of 10^{15} cm^{-3} K, the mass density at 10^5 K is about 2×10^{-14} g cm^{-3}, which for a most probable velocity of 16 km s^{-1} gives an energy density of about 0.12 erg cm^{-3}. By comparison, for a gas pressure of 0.2 dyn cm^{-2}, the internal energy of the plasma, $P/(\gamma - 1)$, is 0.3 erg cm^{-3}. Thus the motions represent a significant amount of energy in these layers of the atmosphere.

Although the nature of the nonthermal broadening is uncertain, it has often been interpreted as the result of the passage of acoustic or MHD waves (e.g., Boland et al., 1975; Mariska et al., 1978; Athay and White, 1978; Bruner and McWhirter, 1979). Following Boland et al. (1975), the energy flux due to waves is

$$\phi = 2EC, \tag{5.15}$$

where C is the propagation velocity of the mode. For acoustic modes, C is the sound speed and for the energy flux we have

$$\phi_S = 3\rho v_{\text{rms}}^2 \left(\frac{5}{3}\frac{P}{\rho}\right)^{1/2}, \tag{5.16}$$

where P is the gas pressure. For MHD fast mode propagation, C is the Alfvén speed, and

$$\phi_F = 3\rho v_{\rm rms}^2 \frac{B}{(4\pi\rho)^{1/2}}, \tag{5.17}$$

where B is the magnetic field strength. Using the perfect gas law, we can eliminate the mass density from equations (5.16) and (5.17), giving

$$\phi_S = 3\left(\frac{5}{3}\frac{\mu m_p}{k}\right)^{1/2} v_{\rm rms}^2 \frac{P}{T^{1/2}} \tag{5.18}$$

and

$$\phi_F = 3\left(\frac{1}{4\pi}\frac{\mu m_p}{k}\right)^{1/2} v_{\rm rms}^2 \left(\frac{P}{T}\right)^{1/2} B. \tag{5.19}$$

Equation (5.15) is based on the assumption that the energy density E is due to a sinusoidal wave propagating in a particular direction. The extension to isotropic nonthermal broadening in equations (5.16) and (5.17) is uncertain. If only a vertical mode is responsible for the energy flux, then the derived fluxes should be reduced by a factor of 3. Since the structures in the transition region are clearly quite complex, the actual factor is probably height dependent. If it is independent of temperature, then the functional form of equations (5.18) and (5.19) will not change.

Because of the small extent of the transition region compared with the corona, many simple models for its structure postulate that the wave flux through it should be roughly constant, or perhaps decrease slightly as some energy is lost. If ϕ_S is constant throughout the transition region, then $v_{\rm rms}$ should be proportional to $T^{1/4}$. Figure 5.4 shows the nonthermal broadening data along with the predicted value of ξ as a function of temperature determined from equation (5.18) with ϕ_S constant. This constant acoustic flux line has been adjusted to fit the data at temperatures near 10^5 K. For a pressure of 0.2 dyn cm^{-2}, the constant flux line has $\phi_S = 1.3 \times 10^6$ erg cm^{-2} s^{-1}. This should be considered an upper limit. If we assume that only a vertical mode is responsible for the broadening, the flux would be reduced to 4.3×10^5 erg cm^{-2} s^{-1}.

Because the magnetic field is an unknown parameter, the situation for the MHD case is more complicated. For a constant pressure, different values of ϕ_F and B can be chosen that give values of $v_{\rm rms}$ consistent with the data. For example, for a pressure of 0.2 dyn cm^{-2}, $\phi_F = 6.5 \times 10^6$ erg cm^{-2} s^{-1} and $B = 10$ G results in the curve plotted in the figure.

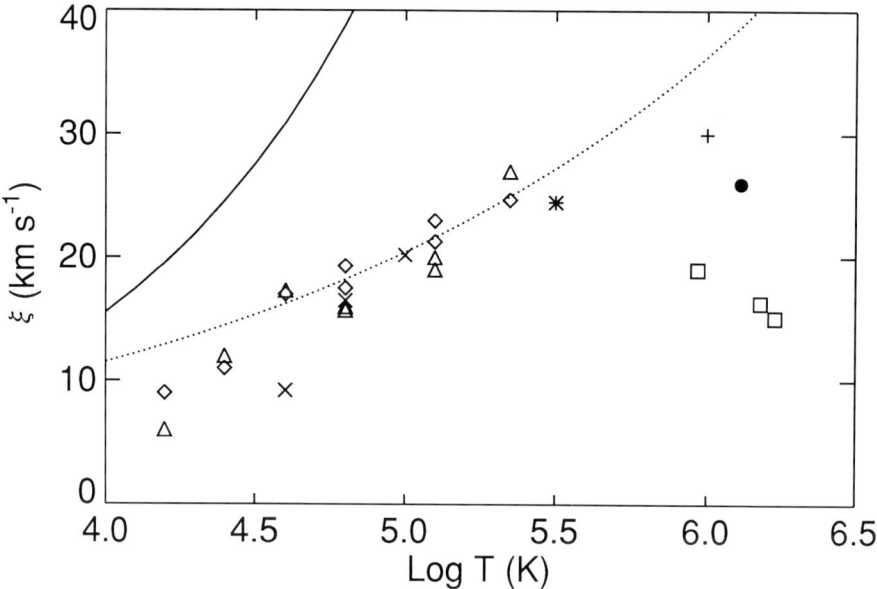

Fig. 5.4. The temperature dependence of the expected nonthermal broadening (dotted line) for a constant acoustic wave flux plotted on the observations shown in Figure 5.2.

Comparison of the curve with the measured nonthermal broadening shows that the data are consistent with a constant acoustic flux in the temperature range from about $40,000$ to 3×10^5 K. Below $40,000$ K, the observations fall below the constant flux line. For acoustic modes this discrepancy suggests a decrease in ϕ_S. For MHD modes it is possible to keep ϕ_F constant by increasing B. Above 3×10^5 K, the data again fall below the curve, suggesting changes in the fluxes or the magnetic field.

A temperature of $40,000$ K also marks the location at which the contrast in emission between network and cells exhibits a sharp decrease from at least 5 to at most about 3 (Feldman et al., 1976b). If the wave interpretation is correct, this would suggest that a different set of UV emitting structures might be responsible for the broadening at lower temperatures. A second explanation, first suggested by Boland et al. (1975) is that a change in area factor of the emitting material is the cause. A factor of 2 change in the area between $20,000$ and 10^5 K would bring the curve into agreement with the data.

Bruner and McWhirter (1979) performed a similar analysis on the C IV 1548 Å line. Using simple model atmospheres, rather than the constant pressure assumption used by Mariska *et al.* (1978), they found that the acoustic flux must be less than 3×10^5 erg cm^{-2} s^{-1} and that for a 10 G field the flux in Alfvén waves can easily exceed 2×10^6 erg cm^{-2} s^{-1} and still not produce all the observed broadening. Their values are lower than those quoted by Mariska *et al.* (1978) because they assumed that they were observing a wave propagating in only one direction, rather than isotropically, leading to a factor of 3 difference between their results and those of Mariska *et al.*

If waves are producing the observed broadening, then observations in a single emission line at increasing heights above the limb also should show an increase in line width. This is due to the drop in pressure with height in the corona. Hassler *et al.* (1990) have observed the increase in line width with height above the limb in the resonance lines of Mg X at 609 and 629 Å. Their preliminary results show that the observed increase in line width is consistent with the presence of hydromagnetic waves in the corona. For a 1 G magnetic field, they estimate an energy flux of 9×10^4 erg cm^{-2} s^{-1}. For a 5 G field the flux is 4.3×10^5 erg cm^{-2} s^{-1}. Note that the value of the energy flux for a magnetic field strength of 1 G is lower than the estimate for the same field strength made by Mariska *et al.* (1978). While some of this is due to the lower electron pressure used by Hassler *et al.* (1990), the difference is large enough to suggest that if we are seeing MHD waves, some wave energy may have been dissipated between the transition region and the corona. This would provide a natural explanation for the turnover in the nonthermal broadening that appears to take place at coronal temperatures in the limb observations.

These estimates are, for the most part, independent of the details of the atmospheric structure. They simply require estimates of the temperature and density at the observed location along with some guess of the value of the magnetic field strength in the transition region and corona. There are of course the implicit assumptions that the transition region and corona are laminar and that the transition region is at constant pressure.

Some attempts have been made to predict what the emission line profile will look like in the presence of sound waves. Byerley *et al.*

(1978) considered the effect of sound waves on the optically-thin 1032 Å O VI line emitted from a low-density plasma, including the case of a constant pressure plasma with a temperature gradient. They found that when the emission is averaged over times that are long compared with the wave period, the line profiles are asymmetric. In all cases the center of gravity of the profile shifted to the blue side of the rest wavelength. Profiles calculated for a single direction of wave propagation differ dramatically from a Gaussian shape. Those for which they averaged a range of propagation directions tended to be closer to Gaussian in shape, but were still noticeably asymmetric.

Bruner and McWhirter (1979) performed the same kind of calculation for the C IV 1548 Å emission line. They assumed an isothermal acoustic wave with a Mach number of 0.5 derived from the observed nonthermal broadening. They also found that the predicted profiles for a single direction of propagation differed markedly from a Gaussian shape. When they considered a distribution of propagation directions, the profile was more nearly Gaussian. At the half-intensity point, however, it still differed from a Gaussian by about 20%, more than an order of magnitude larger than the observed profiles differ from a Gaussian.

These line profile calculations are for acoustic waves only. Alfvén waves will not show an asymmetry in the predicted line profile. Thus an Alfvén wave interpretation of the nonthermal broadening is not ruled out. Moreover, if the transition region contained acoustic waves propagating in all directions, not just over the outward directed hemisphere, then the predicted line profiles would show no asymmetry (Bruner and McWhirter, 1979). For open magnetic field geometries, this would imply no net upward flux of energy. For a closed geometry, however, such as a loop, there could be upward propagating waves in both legs that then went on to become downward propagating waves in the opposite legs.

Nonthermal broadening also can be interpreted as a manifestation of turbulence in the emitting plasma. Hollweg (1984, 1985) has argued that Alfvén waves can drive a Kolmogoroff turbulent cascade. This turbulent cascade is a possible heating mechanism for the entire chromosphere-transition region-corona system. The observed nonthermal broadening in the chromosphere and the corona is consistent with this picture, but more detailed theoretical work is required to develop the idea fully.

Interpreting the observed nonthermal broadening remains an unsolved challenge of transition region physics. It is clear from the data, however, that the energy represented by the nonthermal broadening is substantial and must play some role in the flow of energy in the outer layers of the solar atmosphere. Any successful theory for the structure and energy balance in these layers must in some way account for these data.

5.3 Doppler Shifts

When motions with a size scale that is larger than or comparable to the instrumental spatial resolution are present, then the entire line profile will be shifted in wavelength. Careful measurements of these Doppler shifts thus provide some information on the large-scale dynamics of the plasma in the transition region. Absolute wavelength shifts are difficult to measure, however, because until recently none of the solar rocket or satellite experiments flown included an absolute wavelength standard. Most of the measurements have been made either relative to cool chromospheric emission lines or relative to an average defined by many individual Doppler shift measurements.

5.3.1 Observations

While there have been only a few measurements of Doppler shifts relative to an absolute reference wavelength, wavelength shifts relative to nearby emission lines of neutral, and in some cases singly ionized, elements are almost as useful. Without an absolute wavelength reference, there will always be some uncertainty about the absolute value of the flow velocities obtained. Generally, however, away from chromospheric jets, the emission lines of neutral atoms show much smaller variations in relative Doppler shift than those seen in transition region lines (Dere et al., 1984). Thus the relative measurement can be made quite precisely. For example, Dere et al. (1984) used several Si I lines to establish a wavelength scale and found that they could measure Doppler shifts relative to that standard with an accuracy of about 1.7 km s^{-1}.

Table 5.1 summarizes the published measurements of Doppler

Table 5.1. *Transition region Doppler shifts.*

Ion	λ (Å)	T (10^5K)	Velocity (km s^{-1})	Reference
Si IV	1394, 1403	0.8	9.2	Doschek et al. (1976a)
C IV	1548, 1551	1.0	9.3	Doschek et al. (1976a)
			8.0	Roussel-Dupré and Shine (1982)
			7.0*	Dere et al. (1984)
			10.0	Dere et al. (1989b)
			7.5	Rottman et al. (1990)
			7.7	Hassler et al. (1991)
O IV	1401	1.3	9.9	Doschek et al. (1976a)
N V	1239, 1243	1.8	8.6	Doschek et al. (1976a)
O V	1218	2.2	0.0 ± 8	Doschek et al. (1976a)
Ne VII	465	5.0	0.0 ± 9	Mariska and Dowdy (1991)
Ne VIII	770	8.0	0.0 ± 4	Hassler et al. (1991)

*Corrected for center to limb projection effects.

shifts made relative to chromospheric emission lines, and a few made using other techniques. Except for the C IV resonance lines, most of the published measurements are those made by Doschek et al. (1976a) using data from the NRL spectrograph on *Skylab*. The values listed in the table from that paper are the average of all the locations that showed a measurable Doppler shift. These measurements, the first to show that Doppler shifts were present in the quiet solar transition region, had a spatial resolution of 2 arc sec × 60 arc sec. They showed that when Doppler shifts were present, they were always redshifts with peak velocities of less than 15 km s^{-1}. Moreover, from the small amount of data available, it appeared that the redshifts were confined to bright emitting regions and were thus primarily network phenomena. Doschek et al. (1976a) found Doppler shifts in all of the transition region lines they examined except the O V 1218 Å emission line, which forms at a temperature of 2.2×10^5 K. This line, however, is in the wing of the Lyman α line and thus the Doppler shift measurements are difficult to make.

While it is tempting to use the Doschek *et al.* (1976*a*) measurements to infer differences between network and cell regions and between quiet Sun and coronal holes, it is important to note that they only measured spectra at 18 locations, some of which showed no Doppler shifts at all. Moreover, their measurements were with an instrument with a spatial resolution of 2 arc sec × 60 arc sec. That resolution is sufficiently coarse to make it difficult to distinguish between network and cell. If the entire slit was placed on a network element, as was attempted in the data set, it is not clear that the sizable network segment would represent typical network elements.

To within the accuracy of the measurements, the Doschek *et al.* (1976*a*) data show no differences in the average redshift among the lines for which they observed a redshift. Thus the data suggest that at low spatial resolution and for temperatures of formation between 80,000 and 2×10^5 K, the redshift is roughly constant with an average value of about 9 km s^{-1}. Below a temperature of formation of 80,000 K, their data suggest that the redshift is near zero.

Roussel-Dupré and Shine (1982) made the only other low spatial resolution Doppler shift measurements relative to chromospheric lines. The value of 8.0 km s^{-1} listed in the table for the C IV resonance lines is the average of their measurements. The data they obtained had a spatial resolution of either 3 arc sec × 900 arc sec or 3 arc sec × 20 arc sec, and the result has a quoted standard deviation of about 0.5 km s^{-1}. These data were also time averaged.

Roussel-Dupré and Shine (1982) also used pairs of spectral scans at 3 arc sec × 900 arc sec resolution taken at disk center and at the limb to determine Doppler shifts. If the redshifts represent radial outflows, then limb measurements should provide a rest wavelength measurement for comparison with the disk center measurement. Using this technique, they found a Doppler shift in the C IV resonance lines of 13.0 km s^{-1} with a standard deviation of 3.8 km s^{-1}, and in the Si IV 1393 Å line of 10.3 km s^{-1} with a standard deviation of 4.1 km s^{-1}. Clearly these measurements are subject to considerably more uncertainty than those made relative to chromospheric lines.

Recently Rottman et al. (1990) have obtained the first measurements of Doppler shifts relative to an absolute wavelength calibration. For observations along a solar diameter with a spatial resolution of 20 arc sec × 20 arc sec, they obtain an average radial downflow in the C IV resonance lines of 7.5 km s^{-1}.

Roussel-Dupré and Shine (1982) were also able to measure the center to limb variations in the C IV 1548 Å line and the Si IV 1393 Å line. They did this by constructing average line profiles for 2.7 arc min × 2.3 arc min areas on the Sun at different radial distances from Sun center. Assuming that the line center position for each raster indicates a radial flow at that offset from Sun center, they then fit all of the data with a linear regression function to obtain the Sun center Doppler shift. The values they obtained were consistent with those determined using chromospheric emission lines, but the errors are considerably larger, roughly 4 km s^{-1}. More important, however, is the fact that the data support the idea that the center to limb variations in the Doppler shift are consistent with a radial outflow of material.

Both the *SMM* UVSP and the NRL HRTS have made observations of Doppler shifts in transition region lines at higher spatial resolution. Only the HRTS C IV resonance line Doppler shift observations, however, have been measured relative to chromospheric lines. The velocity of 7.0 km s^{-1} listed in the table was determined by assuming that all the downflows were radial and then correcting for center to limb projection effects (Klimchuk, 1986). The actual average line-of-sight velocity observed during the HRTS 3 rocket flight was 6.0 km s^{-1} (Dere et al., 1984).

While most observations have been made in the C IV resonance lines, there have been some recent attempts to extend Doppler shift measurements to other transition region emission lines. Athay and Dere (1989) have measured line-of-sight velocities in emission lines of Si III, Si IV, and O IV in addition to the C IV 1550 Å emission line. Those velocities are comparable to those measured by Doschek et al. (1976a). At much higher temperatures, Hassler et al. (1991) have measured velocities in the Ne VIII 770 Å line. Their measurements at 20 arc sec × 20 arc sec resolution yield only a limit on the radial velocity of 0 ± 4 km s^{-1}. Mariska and Dowdy (1991) have also placed a limit of 0±9 km s^{-1} on the Doppler shift in the Ne VII 465 Å emission line.

Observations by Lites *et al.* (1976) at 20 arc sec × 20 arc sec resolution show that the Doppler shifts are persistent. They found that the flows they measured in the Si IV 1394 Å emission line using the *OSO 8* UV spectrometer lasted for at least two spacecraft orbits. This result is also true for the 3 arc sec resolution observations reported by Gebbie *et al.* (1981). They found that the same flow pattern persisted for at least 3 hours. Higher-resolution observations for extended periods at the same position on the Sun are not yet available. Whether this persistence in the redshifts continues for observations with 1 arc sec or better resolution will be a key factor in unraveling the nature of the structures responsible for the flows.

It is important to note that these relationships are all statistical in nature. There are wide variations from point to point on the solar surface. The intensity, velocity, and line width data plotted in Figure 5.5 show this vividly. These data are from a HRTS rocket flight in which the spectrograph slit was positioned with one end near Sun center and the other end just above the south polar limb. The intensities are sums of the 1548 and 1550 Å C IV resonance lines; the velocities and widths are averages of the quantities for the two lines. For the line width, the quantity plotted is the second moment of the line profile

$$\Delta\lambda^2 = I^{-1} \int (\lambda - \lambda_{\text{obs}})^2 I_\lambda \, d\lambda, \tag{5.20}$$

where I is the specific intensity and λ_{obs} is the position of the line center. If the line profile is Gaussian, then $\Delta\lambda^2$ is the Gaussian width, which is just the Doppler width in equation (5.11) divided by $\sqrt{2}$.

While the data in the figure clearly show the predominance of redshifts and even contain some hints of statistical relationships, the point to point fluctuations can be dramatic. Also clear is the fact that there are many data points that show blueshifts. These features are more impulsive in nature and will be discussed in the next section.

5.3.2 Relation to Cell-Network Structure

At a spatial resolution of a few arc seconds or better, it is possible to investigate the relation between Doppler shifts and the cell-network structuring of the transition region. This is usually

134 *High-Resolution Spectroscopy*

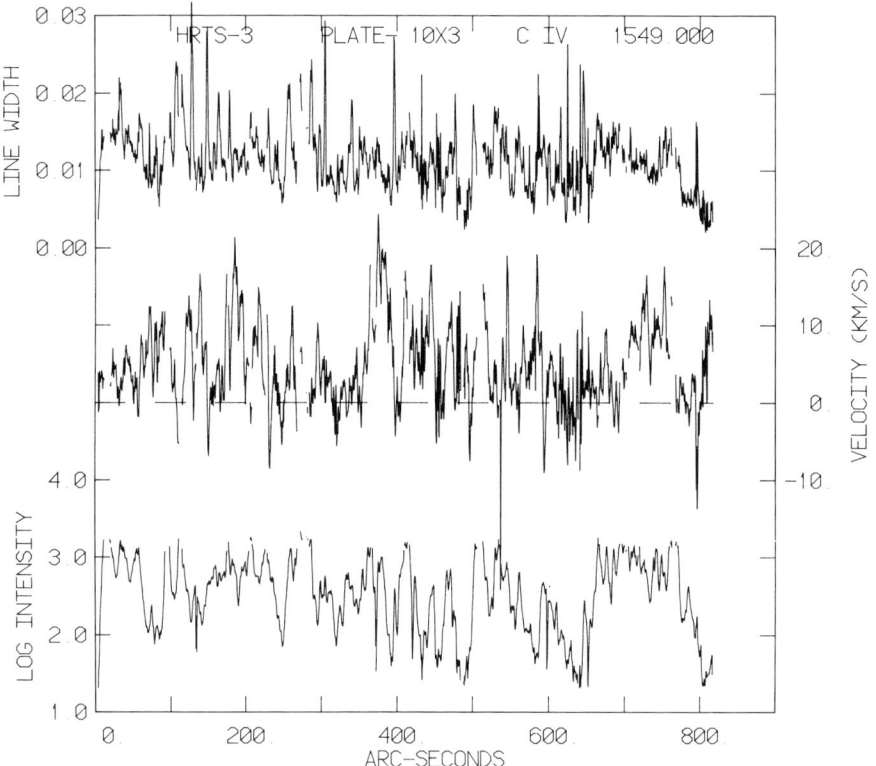

Fig. 5.5. The intensity (erg cm^{-2} s^{-1}), velocity (km s^{-1}), and line width (Å) parameters for the C IV 1548 and 1550 Å emission lines as a function of position along the slit from the limb (at 17 arc sec) to near Sun center (from Dere et al., 1984).

done by examining the relation between the Doppler shift and line intensity, since the network is brighter than cell centers. Gebbie et al. (1981) examined this relation using *SMM* data with a spatial resolution of 3 arc sec. Their velocity measurements were not made relative to cool chromospheric lines. Instead for each 120 arc sec × 120 arc sec raster they measured, they took as the zero for the velocity scale the spatial average of the velocities in the raster. Their data at disk center showed that regions of relative downflow tended to be correlated with regions of bright network emission. Regions of relative upflow tended to occur more often in darker areas.

This correlation is not, however, perfect. When Gebbie et al.

(1981) plotted the C IV 1548 Å data as a velocity versus intensity scatter diagram, they found that the brightest data points formed a distinct branch with a smaller velocity amplitude than the values found for most of the data. The bright data points in this branch were in fact associated with just two bright regions of emission in their data set. Moreover, the distinct lower branch in their scatter diagram was most prominent in data that they had averaged to a spatial resolution of 9 arc sec × 9 arc sec from 120 arc sec × 15 arc sec rasters. Additional observations at higher spatial resolution might fill in the region between the lower velocity amplitude branch and the remainder of the data, reducing the apparent correlation between velocity and intensity.

Other observations do not show this correlation. Athay et al. (1983) examined Doppler shifts in quiet solar regions observed at 3 arc sec resolution with the UVSP on *SMM*. Scatter plots of relative Doppler shift in the C IV 1548 Å line versus intensity do not show the correlation seen by Gebbie et al. (1981). Individual data sets show the branching noted by Gebbie et al., but each data set has the low-intensity branch at a different intensity level. Thus, when several data sets are combined, the effect is to fill in the region between the high-intensity branch and the low-intensity branch. One feature that is present in both analyses is the tendency for the largest redshifts to be associated with intermediate intensities.

While the scatter plots show that there is not a perfect correlation between intensity and Doppler shift, Athay et al. noted that visual comparison of contour plots of velocity and intensity clearly shows that the network is predominately redshifted. Dere et al. (1984) also found no correlation between the Doppler shift in the C IV resonance lines observed at 1 arc sec resolution and the intensity. They also noted, however, that a correlation and branches could be found in subsets of their data. The Dere et al. (1984) data do not even clearly show a tendency for the areas of highest redshift to be of intermediate intensity. There is, however, the same tendency for bright regions to be redshifted as found by Athay et al. (1983). Thus the available observations do suggest that flows are most often associated with the network, but not necessarily the brightest network areas.

5.3.3 Relation to Line Widths

Uncertainty also exists about the relationship between Doppler shifts and line widths. Athay *et al.* (1983) found a correlation between these two quantities, with larger redshifts corresponding to larger line widths. This correlation they argue could be due to a mixture of emitting sources partially filling the observing aperture. This mixture of sources with differing Doppler shifts would then result in a broadened emission line and produce the observed correlation. If this argument is correct, then higher spatial resolution observations should not exhibit a correlation, which is the case. In 1 arc sec resolution data, Dere *et al.* (1984) found no correlation between Doppler shift and line width. Dere *et al.* found, however, that the Doppler shifts localize to regions with sizes in the 2–3 arc sec range. Thus it is not clear that changing the resolution from 3 to 1 arc sec should remove the correlation. If the actual individual emitting structures are much smaller than even 1 arc sec, then both sets of data should show the correlation.

5.3.4 Center to Limb Behavior

While studies of the magnitude of the nonthermal velocity ξ provide some estimates of the magnitude of unresolved motions at the temperature of line formation, studies of the velocity and its fluctuations as a function of position can provide additional information on the nature of the flows. Near Sun center, line-of-sight velocity measurements show the average vertical component of the flows, while near the limb the line-of-sight velocity shows the horizontal component.

At the highest currently available spatial resolution, some of the average properties observed in the lower-resolution data become lost in the small-scale structural details. Thus, while Roussel-Dupré and Shine (1982) found a correlation between redshifts and the cosine of the solar latitude, redshift measurements in a 1 arc sec × 800 arc sec region of the Sun from near Sun center to the south pole showed no cosine latitude dependence (Dere *et al.*, 1984). Another data set for a 0.5 arc sec × 920 arc sec region between Sun center and the southeast limb shows the same effect (Athay and Dere, 1989). Plotted at the full resolution of the observations, the large point-to-point fluctuations obscure any trends that are present. Running averages over 50 arc sec intervals, however, clearly suggest that the line-of-sight velocity in the C IV 1550 Å line decreases from the center of the disk to the limb. This shows that the small-scale velocity fields seen at high spatial resolution have many orientations. Only as a large-scale average do the velocities have the cosine latitude dependence expected for a predominately radial flow.

A given set of C IV Doppler shift measurements provides not only a mean value for the Doppler shift, but also a standard deviation for the mean. A few investigators have noted a tendency for this quantity to increase in observations made close to the limb (e.g., Gebbie et al., 1981; Athay et al., 1983; Klimchuk, 1987). On the other hand Dere et al. (1984) and Athay and Dere (1989) found no evidence for any center to limb change in the root-mean-square (rms) fluctuations in series of C IV velocity measurements. The negative results are based on HRTS measurements, which, while they have 1 arc sec spatial resolution, only cover a small portion of the Sun. The earlier measurements contain both quiet Sun and regions of activity, which makes it difficult to draw firm conclusions from them about the quiet Sun.

Because the Doppler shift data measure the line-of-sight component of the velocity, larger rms fluctuations in the measurements near the limb imply that the horizontal components of the motion vary more than the vertical component. Until the observational picture becomes clearer, however, it is premature to draw that conclusion.

5.3.5 Mass Balance

Although the downflow velocities are small when compared with the local sound speed or the much higher velocities associated with more dynamic events, such as flares, they are still vital to the overall mass and energy balance of the transition region. A downflow velocity of 8 km s^{-1} combined with a typical electron density of 10^{10} cm^{-3} gives a local downward electron flux of 8×10^{15} cm^{-2} s^{-1}. Multiplying this number by the surface area of the Sun then gives a downward total electron flux. This number represents an upper limit, however, since not every point on the surface has a downflow.

Correcting for the area covered by the downflowing plasma is difficult. On the one hand the Doppler shift measurements show that almost every point has a downflow associated with it, suggesting a small correction factor. On the other hand attempts to determine the fraction of the spectrograph slit that is actually filled with plasma emitting in the C IV resonance lines produce filling factors that range from 0.16 to 10^{-5}, depending on what assumptions are made about the geometry of the emitting structures (Feldman et al., 1979; Dere et al., 1984; 1987). Thus the actual downward electron flux could range from 1.3×10^{15} to 8×10^{10} cm^{-2} s^{-1}. Without upflowing material, a downward electron flux near the upper value would drain the entire corona of material in a few thousand seconds. Moreover, additional mass is constantly leaving the corona via the solar wind.

It is generally believed that spicules are the source of the material needed to sustain the corona (e.g., Beckers, 1972). In fact until the discovery of downflows at transition region temperatures, there was a large discrepancy between the measured particle flux in the solar wind and the estimated coronal inflow due to spicules, with the inflow being roughly two orders of magnitude too large.

Estimating the mass inflow to the corona due to spicules is probably as uncertain as estimating the downward mass flux due to the flows seen in UV emission lines. Beckers (1972) estimated a proton number density of about 6×10^{10} cm^{-3}, upflow velocities of 20 km s^{-1}, and a fractional surface area covered by spicules of about 0.1. Assuming the electron number density is roughly equal to the proton number density, this yields a coronal upflow due to spicules of about 1.2×10^{15} cm^{-2} s^{-1}, which is comparable to the larger estimates of the downward mass flux due to the flow seen in UV emission lines. This number represents the upward mass flux at a height of about 3000 km above the limb. At greater heights the flux decreases. For example at a height of 6000 km above the limb, Pneuman and Kopp (1978) estimate an average upward electron flux in spicules of about 6×10^{14} cm^{-2} s^{-1}, while at 8000 km above the limb, the flux is close to 10^{14} cm^{-2} s^{-1}.

The upflowing spicule material is at a temperature of about 10^4 K (Beckers, 1972). No similar large-scale downflow of material at 10^4 K that could be interpreted as the returning spicular material has been observed. In addition, no large-scale upflow of material at 10^5 K has been observed in the UV. Thus the natural conclusion is that the downflows observed at 10^5 K in the UV are produced by the return to the chromosphere of the spicular material observed flowing upward at 10^4 K. To the extent that we can believe the estimates of the area occupied by the upflowing material, this suggests that the area occupied by the downflowing material must be close to 0.16, near the large end of the range of suggested values.

How the spicular material evolves from a 10^4 K, 6×10^{10} cm^{-3} upflowing plasma to a 10^5 K, 10^{10} cm^{-3} downflowing plasma is still unclear. Current observations do not shed any additional light on the subject, and we defer a discussion of possible physical models for the process until later. What is clear from the observations is that a vast circulation system of some kind operates between the chromosphere and the corona. Some coronal material escapes into the solar wind, but 99% of it returns to the chromosphere.

5.3.6 Energy Balance

This returning material not only removes mass from the corona, it also removes energy. A steadily moving fluid carries energy in the form of a kinetic energy flux $\frac{1}{2}\rho v^3$ and an enthalpy flux $\gamma P v/(\gamma - 1)$. The enthalpy flux is composed of two parts, the flux of internal energy transported by the flow $Pv/(\gamma-1)$, and the work done by the pressure force Pv. At transition region densities and velocities, the kinetic energy flux is small, but the enthalpy flux can be substantial. Taking a pressure of 0.2 dyn cm^{-2}, a velocity of 8 km s^{-1}, and a ratio of specific heats of 5/3, we have an enthalpy flux of 4×10^5 erg cm^{-2} s^{-1}. Radiation is the primary energy loss mechanism from the lower transition region. Kopp (1972) estimated that the total rate of radiative energy loss from the transition region is about 1.9×10^5 erg cm^{-2} s^{-1}. Thus the energy supplied to the lower transition region by mass downflows is comparable to the energy required to maintain the observed radiative loss.

As with determinations of the mass flux, determinations of the actual enthalpy flux are uncertain because of our lack of a detailed picture of the geometrical structure of the transition region. Using the same range of possible filling factors that we used for estimating the downward mass flux gives energy fluxes that range from 64,000 to 4 erg cm^{-2} s^{-1}. Thus, only if the fractional surface area covered by the downflowing material is near the upper range of the values that have been suggested, will the energy carried by the flowing material be important for the global energy balance.

In addition, the estimate of the radiative energy losses from the transition region is a global average. The high-resolution observations that we discussed earlier show that most of the emission is concentrated in the network. In local areas on the Sun then, the radiative flux can be much larger. Clearly more detailed observations of flows in individual structural elements will be required to sort out the importance of the energy and mass carried by the flowing plasma. The key point that the observations imply is that, even at small velocities, flows of transition region material transport considerable amounts of mass and energy.

A flow velocity of 8 km s^{-1} in the C IV resonance lines represents a wavelength shift of only 0.04 Å. Even at 1 arc sec spatial resolution, the downflowing features seen in transition region emission lines have Gaussian line profiles (Dere et al., 1984). For the C IV resonance lines, a typical full width at half-maximum is near 0.2 Å. This means that a substantial amount of the observed emission in the line profiles is coming from the short wavelength side of the rest wavelength. Thus the observed Doppler shifts do not unambiguously show that nearly all the material in the transition region at 10^5 K is flowing downward. Instead they permit several possible interpretations. If, even at 1 arc sec resolution, there are several independent emitting regions in the spectrograph field of view, the results could mean that more regions are moving down than are moving up. A second possibility is that there are several regions in the field of view with the same number moving in each direction, but the downflowing regions are for some reason brighter.

Doschek et al. (1976a) first suggested these two explanations based on observations made with a 2 arc sec × 60 arc sec spectrograph slit. An aperture that large could easily permit a sufficient number of small-scale structures to produce a Gaussian line profile and yet still have little or no actual downflows of material in the observed region. This argument is more difficult to make with the 1 arc sec resolution observations. Thus the third possibility is that the observed broadening of the emission lines is intrinsic to the plasma at the smallest spatial scales and the Doppler shifts do indicate that virtually all the emitting plasma in the spectrograph aperture is flowing downward. The major problem with this explanation is the lack of any observed upflow at 10^5 K.

At high spatial resolution, the measured Doppler shifts show large point-to-point fluctuations (see, for example, Figure 5.5). Dere (1989) has used this information for the C IV resonance lines to compute the velocity power spectrum for wavenumbers $k < 0.008$ km^{-1}. He then extrapolated this power spectrum to larger wavenumbers to match the observed unresolved nonthermal velocities, arriving at limits on the exponent α of the extrapolated power, which he assumed varied as $k^{-\alpha}$. Dere found that $0 < \alpha < 1.0$, and that significant power was present at high wavenumbers. These values of α are smaller than those predicted for the incrtial range of a turbulent fluid, suggesting that a turbulent cascade is not maintaining the transition region as suggested by Hollweg (1984). Further power spectrum analysis of high spatial resolution data should shed more light on the role these small-scale velocity fluctuations play in heating the transition region.

5.4 Non-Gaussian Line Profiles

While low spatial resolution observations show redshifted emission, high-resolution observations such as those in Figure 5.6 also show blueshifted features and non-Gaussian line profiles. Unlike the features that produce primarily redshifted emission, these features tend to vary on relatively short time scales. Brueckner and Bartoe (1983) have classified them into two groups: turbulent events and jets. Dere *et al.* (1989a) have suggested that the name explosive events is more appropriate.

5.4.1 *Explosive Events*

Turbulent events were first identified in high-resolution rocket spectra (Brueckner and Bartoe, 1983). The spectra in Figure 5.6 display several of them. In the spectra they are characterized by non-Gaussian enhancements of the long- and short-wavelength wings of the line profiles in regions roughly 2 arc sec or less in size. Analyses of events observed in the NRL HRTS-3 rocket flight provide most of the published data on turbulent events. Brueckner and Bartoe (1983) examined 145 events and summarized their general characteristics. Dere *et al.* (1989a) performed a more detailed

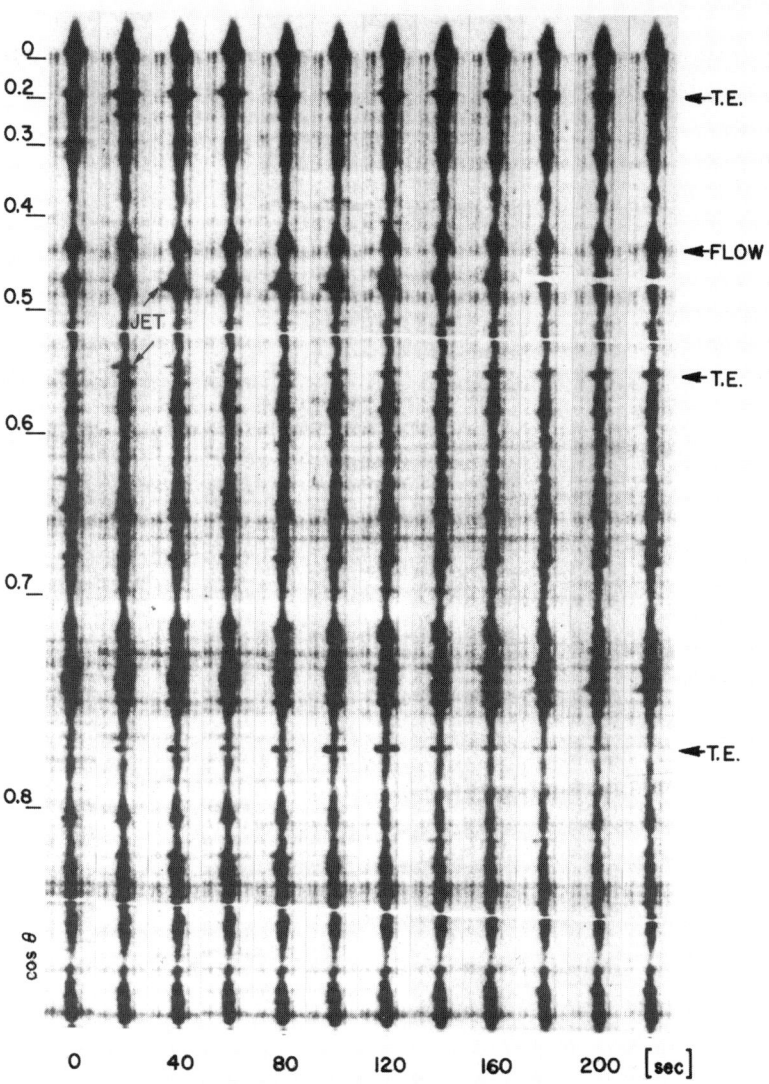

Fig. 5.6. Time sequence of C IV spectra showing dynamic fine structure in the transition region. In each spectrum, short wavelengths are to the left and the solar limb is at the top. The angle θ is the angle between the line of sight and the normal to the solar surface (courtesy G. E. Brueckner, NRL).

study of 82 events from the same rocket flight. Cook et al. (1988) provided a preliminary summary of the *Spacelab 2* observations.

Brueckner and Bartoe (1983) found that explosive events could be detected in emission lines formed between 20,000 (C II 1334 Å, 1336 Å) and 2×10^5 K (N V 1239 Å). They were not visible in emission lines of Si II, C I, and O I, which form at lower temperatures, and there are no useful higher-temperature lines in the HRTS data. A crude emission measure analysis for a large event at 150 km s^{-1} to the short-wavelength side of the rest wavelength yielded values of the differential emission measure at 80,000 and 10^5 K that were the same as those seen in ordinary quiet-Sun downflows. At 20,000 K, however, the emission measure was an order of magnitude smaller than in a typical downflow, while at 2×10^5 K it was a factor of 2 smaller.

Observations in the C IV resonance lines provide all the detailed information on explosive events. In these lines Dere et al. (1989a) found that the average extent of the events along the slit was about 15,000 km, or about 2 arc sec. Often the positions of the blue and red wings of a single event were displaced along the slit, also indicating a size of around 15,000 km. They measured average maximum Doppler shifts of 108 km s^{-1} in the blueshifted component and 114 km s^{-1} in the redshifted component. Brueckner and Bartoe (1983) observed maximum velocities that ranged from ± 50 km s^{-1} to ± 250 km s^{-1}. For velocities below 50 km s^{-1}, the events become increasingly harder to detect because the emission merges with the stronger background line profile. Since the Doppler shift measures the line-of-sight velocity component, large numbers of measurements made at different distances from the limb should reveal any tendency for the actual motions to be either radial or tangential. For the 82 explosive events Dere et al. (1989a) examined, the maximum line-of-sight velocity was independent of viewing angle. Thus explosive events produce roughly equal flows in all directions.

Individual explosive events evolve on time scales ranging from 20 to 200 s. Since many events evolve in a complicated manner, it is not always possible to assign a single time scale. Dere et al. (1989a) concentrated on the time scale for the short-lived high-velocity component, ignoring the low-velocity event that often accompanies it. They found an average lifetime of 60 s for the 82 events they studied. Brueckner and Bartoe (1983), on the other hand, found an average lifetime for the 145 events they studied of 40 s. The distribution of lifetimes they show actually peaks at 20 s, the observing cadence of the data. Using *Spacelab 2* data, Cook et al. (1988) were only able to obtain an upper limit of 90 s for the lifetime. Additional high time resolution observations with total durations of several minutes should result in significant improvements in these estimates.

Along the spectrograph slit the HRTS rocket had a spatial resolution of roughly 1 arc sec, about 725 km. A feature with a lifetime of 100 s and an average velocity of 50 km s^{-1} should move 5000 km, almost 7 arc sec, during its lifetime. Thus any proper motion along the slit by the emitting plasma in an explosive event should be easily visible. Of the 82 events they examined, Dere et al. (1989a) only saw one case in which there was evidence for apparent velocities along the slit. This result is puzzling, since the lack of a latitudinal dependence in the measured velocities suggests that the velocities are isotropic. It may be the result of the limited temperature coverage of the HRTS instrument. If the plasma is heated as it accelerates, then it would cease to be visible in the C IV resonance lines before it left the field of view of the spectrograph (Dere et al., 1989a). For the events to last an average of 60 s, this means that new plasma must be continually accelerated and heated. Thus the mechanism for creating the explosive events may be more continuous than an explosion. An alternative explanation is that small magnetic structures confine the plasma and it is in some kind of highly turbulent state. Additional high-resolution observations, particularly in emission lines formed at higher temperatures, will be required to understand further this aspect of these events.

With no detectable proper motion, it is difficult to measure unambiguously the acceleration of the plasma in an explosive event. Dere et al. (1989a) estimated a value of about 5 km s^{-2} based on the observation that an event can appear with a velocity of 100 km s^{-1} in the 20 s between spectra. Since most of the observations are confined to the C IV resonance lines, it is difficult to know the history of the plasma. Thus acceleration estimates are probably misleading.

Although the spatial coverage of the explosive event observations is limited, it is possible to reach some conclusions about their spatial distribution on the Sun. Using their limited set of data, Dere et al.

(1989a) found that there was some tendency for explosive events to be correlated with the network as indicated by bright areas seen in Fe II and C IV emission lines. They also noted some tendency for the events to occur at midlatitudes. Brueckner and Bartoe (1983) pointed out that smaller-velocity turbulent events are more easily detected in cell interiors because of the absence of strong background emission. Thus the relation between explosive events and other features of the outer layers of the solar atmosphere is still uncertain.

Despite the poor spatial and temporal coverage of the available observations, it is possible to estimate roughly explosive event birthrates. Based on 53 observed events in quiet regions during 220 s of observing, Dere et al. (1989a) estimated a quiet-Sun birthrate of 1×10^{-20} cm^{-2} s^{-1}, which corresponds to a global birthrate of 600 s^{-1}. They also estimated a birthrate of 4×10^{-21} cm^{-2} s^{-1} in coronal holes based on 25 observed events. Brueckner and Bartoe (1983) estimated a global birthrate of 753 s^{-1} based on their count of 145 events in 260 s, with no attempt made to distinguish between quiet Sun, coronal holes, and active regions. Cook et al. (1988) examined the larger rasters obtained with the *Spacelab 2* HRTS in the C IV resonance lines and counted 549 explosive events. From this they deduced a global birthrate of 44 s^{-1}, much smaller than the estimates from the rocket observations. This discrepancy may be in part due to their sampling time of 90 s being larger than the average lifetime of 60 s determined from the 20 s time resolution rocket observations.

When observed in the C IV resonance lines, individual explosive events exhibit a variety of time histories. The only element common to all observed explosive events is strong emission at or near the rest wavelength. Dere et al. (1989a) presented a number of examples to illustrate the variety of line profiles seen, and suggest that they could be divided into three general shapes. First are events in which the profiles decrease in intensity from the central emission core to higher absolute velocities in the line wings. This pattern can be present in either wing or in both. Second are profiles in which the emission beyond the central core is roughly constant over a range of velocities. Finally, there are cases in which the emission away from the central core peaks at some intermediate velocity and then decreases again at larger velocities. In a single explosive event, the

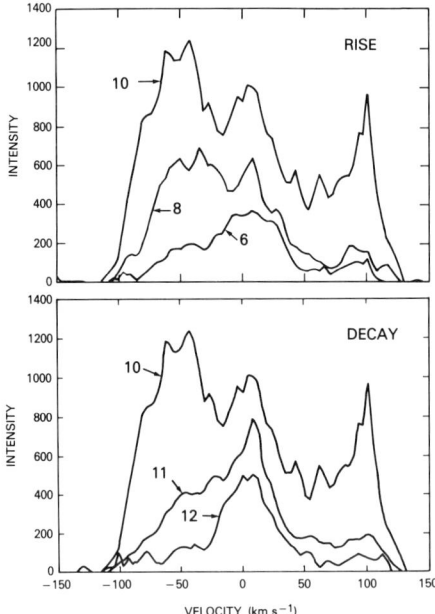

Fig. 5.7. C IV resonance line profiles during the rise phase (top panel) and decay phase (bottom panel) of an explosive event. The time difference between each spectrum in the top panel is 40 s, while the time difference between each spectrum in the bottom panel is 20 s (after Dere et al., 1989a).

line profile can exhibit different shapes at different times. Figure 5.7 shows one example of the time history of the line profile of a typical event in the C IV resonance lines.

Because explosive events are small and have been observed in only a few emission lines, estimates of key physical properties, such as their temperature and density structure and their mass and kinetic energy, are difficult to make. Dere et al. (1989a) used the integrated emission in the red and blue wings of the C IV resonance line profiles to estimate an average volume emission measure $(n_e^2 V)$ at 10^5 K of 4×10^{41} cm^{-3}. This is somewhat smaller than the value of 4×10^{42} cm^{-3} obtained by Brueckner and Bartoe (1983) for a large event. From this emission measure, the mass of the plasma in the event at 10^5 K is given by

$$M = m_H n_e V = m_H \left(n_e^2\right)^{1/2} (fV)^{1/2}. \tag{5.21}$$

Here f is a filling factor to account for the possibility that the apparent emitting volume V may be larger than the actual volume because of unresolved fine structure. Assuming the volume of emitting plasma is a sphere with a diameter equal to the full width at half-maximum of an event along the slit, Dere et al. (1989a) estimated an average mass in the blue wing of 5×10^8 g and in the red wing of 7×10^8 g. Using the average of these values and an average peak velocity of 111 km s^{-1} yields an average kinetic energy of 4×10^{22} erg.

This kinetic energy and the quiet-Sun birthrate of 600 s^{-1} implies a kinetic energy flux of about 3.9×10^2 erg cm^{-2} s^{-1}, which is small compared with the 1.9×10^5 erg cm^{-2} s^{-1} lost by radiation from the transition region (e.g., Kopp, 1972). Thus the kinetic energy in explosive events does not represent a significant energy loss from the transition region. In addition this implies that any heating mechanism that provides for the radiative losses of the transition region also could easily provide enough energy to supply the kinetic energy requirements of the explosive events—at least as a global average.

These estimates are of course subject to large uncertainties. There is, for example, no independent estimate of the density, which would provide an estimate of the filling factor. Using the average volume for an explosive event and the emission measure estimated above yields an electron density at 10^5 K of about 1×10^9 cm^{-3}. Quiet-Sun density-sensitive line ratios typically yield density estimates of about 1×10^{10} cm^{-3} at 10^5 K. Thus it is likely that the filling factor is on the order of 0.1 and the mass and energy are at least a factor of three smaller than the values estimated above. On the other hand, this mass estimate is based only on the emission observed near 10^5 K. Since Brueckner and Bartoe (1983) found significant emission measure in lower-temperature lines also, this would yield higher total masses and kinetic energies.

Porter et al. (1987) have examined sequences of images obtained in the 1548 Å C IV line with the *SMM* UVSP and found events that are similar in behavior to the explosive events seen in the HRTS spectra. These microflares have lifetimes that are similar to those of the events seen in the HRTS data. Many of them also have spatial scales that are similar, although this is more difficult to determine because of the 3 arc sec resolution of the UVSP observations used in the study. Comparing their UV observations with near simultaneous Kitt Peak magnetograms, Porter *et al.* (1987) found that the microflares correspond to the sites of the neutral lines of small magnetic bipoles. Moreover, the stronger events correspond to stronger bipoles. Comparison with near simultaneous He I 10830 Å spectroheliograms further showed that the stronger events correspond to He I dark points, which coincide with X-ray bright points in the corona. Bright points are small magnetic bipoles, often appearing to consist of small loops (e.g., Krieger *et al.*, 1971; Sheeley and Golub, 1979). Thus Porter *et al.* (1987) suggest that explosive events are simply part of the same class of events as X-ray bright points. Many of them are simply too small to produce appreciable X-ray emission.

Although the evidence for a relationship between explosive events and magnetic bipoles is strong, the details remain unclear. Brueckner *et al.* (1988) studied one particularly large explosive event observed at the edge of an active region with the HRTS on *Spacelab 2*. They found that Hα images and magnetograms showed that the location with the broadest profiles corresponded to a region of emerging magnetic flux. This suggests that the events may be driven by changes in the magnetic field, either the emergence of new flux or reconnection in existing magnetic structures.

Recently Porter and Dere (1991) have investigated the relationship between HRTS *Spacelab 2* data on explosive events, a coaligned simultaneous magnetogram, and a coaligned near simultaneous He I 10830 Å spectroheliogram. They found that the explosive events occurred in the magnetic network, but not at the locations of the largest concentrations of magnetic flux. When the explosive events were at the site of a magnetic bipole, it was not the strongest bipole, but rather a less conspicuous one. These observations suggest that the microflares examined by Porter *et al.* (1987) in the UVSP data are different from the explosive events observed with the HRTS. The microflares appear to have lower velocities, are generally brighter than the explosive events, and are associated with the stronger network fields. Thus the explosive events are peculiar in that they are energetic and yet are not associated with the strongest local magnetic field concentrations (Porter and Dere, 1991).

Because explosive events have large peak velocities and are numerous, they represent a potential source of mass for the corona and solar wind. The upward mass flux in explosive events is given by MB, where M is the average mass measured in the blue wing of the C IV resonance lines and B is the birthrate. Taking the quiet-Sun birthrate and the blue wing mass estimate, the upward mass flux in explosive events is about 5×10^{-12} g cm^{-2} s^{-1}. This corresponds to a proton flux of about 3×10^{13} cm^{-2} s^{-1}. In comparison the solar wind proton flux at the Sun is about 1.5×10^{13} cm^{-2} s^{-1} (e.g., Schwenn, 1983; Withbroe, 1988). Clearly explosive events could be important in the mass balance of the solar corona and solar wind. Lack of significant proper motions in the events, however, as well as no confirming observations of explosive upflows in emission lines formed closer to coronal temperatures, suggest that they are not directly the source of the solar wind.

5.4.2 Jets

Jets have a strongly enhanced short-wavelength wing showing rapid velocity growth to as much as 400 km s^{-1}. Brueckner and Bartoe (1983) placed them in a separate class from explosive events because only the blue wing showed enhanced emission and the maximum velocity was much larger than the ± 250 km s^{-1} peak values measured in those features. Often, however, there was also a red component at a smaller velocity. Once the jets reach their peak velocity, they disappear, either because of motion out of the field of view or because of heating. Measurements of the acceleration at the leading edges of four jets showed roughly a constant value of 5 km s^{-2}.

Like explosive events, jets are most easily seen in the C IV resonance lines. Strong ones, however, have been detected in lines ranging from the Lyman α line in the upper chromosphere to the O V 1218 Å line, which forms at a temperature of about 2.2×10^5 K. Emission measure estimates show that the general character of the emission measure as a function of temperature is similar to that in explosive events. As the jet evolves, however, the relative amount of material at 2×10^5 K near the upper limit of observable temperatures in the HRTS spectra appears to increase with time relative to the amount of material at 20,000 K, near the lower limit of observable temperatures.

Brueckner and Bartoe (1983) have argued that the emission spikes seen at the limb in spectroheliograms made in a wavelength band

Table 5.2. *Average properties of explosive events and spicules.*

Parameter	Explosive Events	Spicules
Peak velocity (km s^{-1})	108	25
Lifetime (s)	60	300–960
Horizontal size (km)	1500	400–1500
Radial size (km)	?	6500–9500
Quiet-Sun birthrate (s^{-1})	600	3300–1000
n_e (cm^{-3})	$> 10^9$	$3.4–16 \times 10^{10}$
Mass (g)	6×10^8	6×10^{11}
Electron flux (cm^{-2} s^{-1})	3×10^{13}	1.2×10^{15}
Kinetic energy (erg)	4×10^{22}	2×10^{24}
Kinetic energy flux (erg cm^{-2} s^{-1})	4×10^2	5×10^3

centered on the C IV resonance lines correspond to the jets observed on the disk. So far, however, there is no convincing evidence to support this assertion. In fact, only seven individual jets have been detected in three rocket flights. Based on the detailed studies of the properties of many more explosive events (Dere et al., 1989a), it appears that jets are simply examples of explosive events with higher peak velocities than the average.

5.4.3 Relation to Bright Points and Spicules

Porter et al. (1987) have already shown that it is likely that the strongest microflares are related to X-ray bright points in the corona. How the small-scale dynamical features of the transition region relate to chromospheric structure is much less clear. Since explosive events show some tendency to occur in regions of network emission, they may be related to network fine structure such as spicules. Table 5.2 compares the average properties of explosive events and those of spicules as summarized by Beckers (1972). Clearly there are few indications that explosive events are spicules seen in UV emission lines. On the other hand, the properties of the explosive events are all consistent with the possibility that they represent a subset of the spicule population.

There is already some observational evidence that suggests that Hα spicules may be related to enhancements in higher-temperature emission lines. Moore et al. (1977) found that EUV macrospicules could be identified with Hα eruptions. These Hα macrospicules have peak velocities of about 100 km s^{-1}, similar to those seen in explosive events. In addition, simultaneous observations in soft X-rays and Hα show that Hα macrospicules, like some explosive events, are related to flaring X-ray bright points. Their lifetimes, however, are 5–15 minutes, much longer than those seen in explosive events. Based on this relationship between Hα and EUV macrospicules and flaring in X-ray bright points, Moore et al. (1977) suggested that smaller microflares produce spicules. Porter et al. (1987) have further suggested that the brightenings they observe in the UV in the quiet network are the microflares which produce spicules. While these same brightenings do not appear to be identical with the explosive events observed with the HRTS, it is likely that they are part of the same class of events.

This picture is appealing. It links phenomena observed from the chromosphere to the corona to magnetic bipoles and the energy release associated with changes in the magnetic field structure. At present, however, the link from macrospicules to spicules is weak. Spicules are just one example of the complex of fine-scale structure observed in the quiet chromospheric network. Based on the crude data presented in the table, there is no strong reason to link them with UV explosive events. Only additional multiwavelength observations will unravel the relationships among chromospheric fine structure, explosive events, and the small scale X-ray features of the corona.

5.5 Temporal Variations

High spectral and temporal resolution observations allow us to examine changes not only in the total line intensity but also in the Doppler shift and line width. These extra parameters provide additional information on any waves that might be present, since we can now determine phase differences between changes in different characteristics of the emission line profile.

An acoustic wave propagating through the transition region will perturb the ambient plasma and betray its presence through changes in the UV emission lines formed there. At any location in the transition region, the upward passage of a linear acoustic wave will produce a small compression in the plasma followed by a small rarefaction. A small upward velocity accompanies the compression and a small downward velocity accompanies the rarefaction. For a downward propagating acoustic wave the compression would be associated with a small downward velocity and the rarefaction would be accompanied by an upward velocity.

Since the intensity in an optically-thin UV emission line depends on the square of the density, the compression and rarefaction of the plasma result in intensity fluctuations, while the velocity signature will appear as a Doppler shift. The small changes in temperature which accompany the passage of a low-amplitude adiabatic acoustic wave also will modulate the line intensity. For transition region lines, such as the C IV resonance lines, however, the intensity modulation due to the density fluctuation is about a factor of 6 larger than the modulation due to temperature fluctuations (Bruner, 1981). Thus the intensity and Doppler shift should be correlated if an acoustic wave is propagating through the transition region. The phase delay between the intensity and upward velocity should be 0° for upward propagating acoustic waves and 180° for downward propagating acoustic waves.

An Alfvén wave, on the other hand, is a transverse movement of the magnetic field lines and is not compressive. The plasma simply moves in the transverse direction with the field lines. Thus, only velocity perturbations would be expected if Alfvén waves are present in the transition region. For a radial magnetic field, the motions at Sun center would be transverse to the line of sight and undetectable as Doppler shifts. At the limb, the motions would be in the line of sight and thus detectable. Thus Alfvén waves would betray themselves as periodic fluctuations in the Doppler shifts near the limb with no such shift at Sun center and no accompanying intensity fluctuations.

Unfortunately, the fine structure in the upper chromosphere and transition region clearly shows that the magnetic field is not always radial. In addition, it is unlikely that only pure Alfvén waves or pure acoustic waves are present. A complicated combination of the two observed at a nonzero angle to the magnetic field direction is more likely to be the case, leading to more complex correlations between the brightness and velocity at any point in the transition region.

Searching for temporal variations in the transition region has been motivated primarily by the search for evidence of the passage of the waves that were thought to be heating the corona. Shortly after the discovery of the true temperature of the corona, Biermann (1946), Schwarzschild (1948), and Schatzman (1949) suggested that acoustic waves generated in the convection zone supplied the energy needed to heat the outer layers of the solar atmosphere. As the waves propagate upward they steepen into shocks and dissipate their energy in the chromosphere and corona. Later, when the nature of the magnetic field structure of the solar atmosphere began to emerge, Osterbrock (1961) examined the generation of Alfvén waves in the convection zone and their upward propagation.

Observers of the solar photosphere have long recognized that atmospheric motions are common. Spectroscopic observations show clear evidence for atmospheric motions in the wiggly line spectra and in the microturbulence required to force agreement between observed and calculated line profiles. The modern study of waves in the solar atmosphere began when Leighton *et al.* (1962) discovered that quasiperiodic oscillations with a period near 5 minutes were a common feature of the quiet solar photosphere. In recent years the study of waves in the solar atmosphere has become a major subfield of solar physics, helioseismology. Stein and Leibacher (1974) summarized much of the early work. More recent reviews include those by Deubner (1981), Leibacher and Stein (1981), Stein and Leibacher (1981), and Deubner and Gough (1984).

We now know that the 5 minute oscillations are standing acoustic waves, most of which are trapped in a subphotospheric cavity (Ulrich, 1970; Ando and Osaki, 1975; Deubner, 1975). Above the photosphere, Ando and Osaki (1975) predicted a second cavity associated with the chromosphere. In the chromosphere, however, magnetic fields and small-scale structure are more important than in the photosphere and simple acoustic waves may not be present.

In the low chromosphere, the observed oscillations have periods in the range from 180 to 240 s (e.g., Deubner, 1981). It is not clear, however, whether the waves are evanescent or propagating. Studies by Mein (1977), Schmieder (1978), and Schmieder and Mein (1980) suggested that there was little energy propagation. Lites and Chipman (1979) on the other hand found that in the low chromosphere the 5 minute oscillations were evanescent, but at frequencies above 5 mHz (200 s periods) there was evidence for propagating waves. The total upward mechanical energy flux in the waves was only 68,000 erg cm^{-2} s^{-1} at a height of about 900 km in the chromosphere. That is about two orders of magnitude below the heating requirements for the overlying layers. In a study of velocity oscillations of a pair of Ca II lines formed in the chromosphere, Lites et al. (1982) found no evidence for vertically propagating acoustic disturbances. They and others have pointed out that these measurements are difficult to make from the ground because the observed lines are optically thick and radiation transfer effects resulting from the velocity fluctuations might be important.

Observational searches for waves in the upper chromosphere and the transition region have been undertaken using UV emission lines. These lines are effectively thin, which removes some of the uncertainty associated with radiative transfer effects. Using data from *OSO 8* for UV lines of Fe II, Si II, and C II, Chipman (1978) found that not all the observations showed periodic oscillations. For those that showed oscillations, the predominant oscillatory period was near 300 s, rather than the 180–240 s oscillations observed lower in the chromosphere and predicted for the chromospheric cavity. Measurements of phase differences between the chromospheric lines and the UV continuum formed near the temperature minimum showed only small differences in the phase of the intensity oscillations. This resulted in derived phase velocities that were much larger than the sound velocity, suggesting that the waves were evanescent.

Athay and White (1979*a, b*) examined an extensive set of time series of emission-line profiles in the Si II 1816.83 and 1817.45 Å lines. They noted that in quiet regions oscillations with periods near 300 s were almost always present. Power spectra calculated from time series of line intensity and Doppler shift measurements were similar in appearance. They contained a broad power maximum in the range from 2.5 to 9 mHz (400–110 s) superposed on a flat noise spectrum that extended to beyond 30 mHz. The location of the broad peak shifted to lower frequencies and narrowed for data with increased line strength. For all the data sets, however, the peak was near a period of 300 s, not the 180–240 s observed from the ground. The flat noise spectrum contained most of the actual observed power. Of this power, Athay and White estimated that less than half was from solar fluctuations. Combining the energies in solar oscillations at all frequencies up to 30 mHz, they estimated a total flux of 10,000 erg cm^{-2} s^{-1}, which is more than an order of magnitude less than the amount required to heat overlying atmospheric layers.

Unlike Chipman (1978), White and Athay (1979) found evidence that the waves with periods near 300 s propagated vertically. In the frequency range from 2 to 9 mHz, they found delays between oscillations in the two lines that were consistent with upward propagation at velocities near the sound velocity of 7 km s^{-1}. Maximum brightness led maximum blueshift by 60°, rather than coinciding with it as expected for a pure acoustic wave. White and Athay suggested that this was the result of the temperature and density structure of the chromosphere.

In the transition region, observations from *OSO 8* in the C IV 1548 Å line and Si IV 1393 Å line have been analyzed for oscillations. For the C IV, Athay and White (1979a) found that low-amplitude aperiodic fluctuations characterized most of the data. Only about 20% of the data sets showed periodic fluctuations with power in the 3–5 mHz region. For this reason average power spectra showed only a systematically higher power level in the 3–5 mHz region instead of a distinct peak. Power spectra constructed using data sets that showed oscillatory behavior showed more power in the 3–5 mHz range, but still no distinct peak because of contributions at lower frequencies for irregular fluctuations that are common on time scales of 10–20 minutes. For the data sets that showed oscillatory behavior in the 3–5 mHz region, the maximum brightness led the maximum blueshift by 60°, the same value they found in the Si II lines. In addition, there was some evidence for a delay of about 300 s between oscillations in the continuum emission formed near the temperature minimum and the C IV resonance line oscillations. This disagreed with Chipman's (1978) conclusion that intensity oscillations in the C IV and Si IV lines preceded temperature minimum oscillations by about 10 s. In the aperiodic events they found that the intensity was correlated with a blueshift.

Athay and White (1979a) interpreted the C IV results as suggesting that sound waves are common at 10^5 K in the transition region, but that often their passage through the upper chromosphere and transition region has destroyed their periodicity. They estimated the total energy flux carried by the periodic and aperiodic sound waves to be about 10,000 erg cm^{-2} s^{-1}, substantially below the heating requirements for the corona.

Bruner (1978) also examined time series obtained with *OSO 8* in the C IV 1548 Å emission line and reached similar conclusions. Oscillations with periods near 300 s were not common and the energy flux observed was insufficient to heat the overlying corona. In a careful reanalysis of the C IV data, Bruner (1981) concluded that, while there was evidence for acoustic waves in the brightness and velocity fluctuations, on the average as much acoustic flux was carried down as was carried up. The resulting upper limit on the acoustic flux available to heat the corona was 300 erg cm^{-2} s^{-1}.

All the observations of the upper chromosphere and transition region show ample evidence for the presence of acoustic waves. In the chromosphere they are common. In the transition region they are much less common. Analyses of the observations in both regions show that there is insufficient energy present in the acoustic disturbances to heat the corona. Beyond that conclusion, the details of how the disturbances evolve in the upper chromosphere and transition region are unclear.

These searches for waves and oscillations were based on observations made at 1–2 arc sec × 20 arc sec spatial resolution with the *OSO 8* spacecraft. Thus, while the spectral resolution of the observations was adequate for detailing intensity and velocity fluctuations, the spatial resolution was clearly much larger than the 3 arc sec or less characteristic size of typical emission elements seen with the higher resolution HRTS experiment (Dere *et al.*, 1987). This adds considerable uncertainty to the interpretation of the time series. Although the primary conclusion that acoustic waves do not heat the corona is unlikely to change, a much clearer picture of the time-dependent behavior of the plasma in the upper chromosphere and transition region will emerge when longer time series of observations with instruments such as the HRTS become available.

Table 5.3. *Ionization and recombination times for carbon.*

Ion	$\tau_I(z)$ (s)			$\tau_R(z)$ (s)		
$\log T =$	4.8	5.0	5.2	4.8	5.0	5.2
C III	1.4+2	4.4	4.5−1	2.4	2.1	2.4
C IV	1.1+4	1.1+2	5.4	3.4	4.0	5.7
C V	...	...	...	6.0+1	8.7+1	1.3+2

5.6 Diagnostic Implications of a Dynamic Transition Region

High-resolution observations show that the temperature structure of the transition region must be complex. Small transition region structures penetrate well into the surrounding 10^6 K corona, while the average thickness is only a few thousand kilometers. Clearly significant temperature gradients are present in the transition region. Probably the steepest temperature gradients are perpendicular to the magnetic field, a direction in which plasma flows are unlikely. Even along the magnetic field, however, the temperature probably changes quite rapidly. Thus for the steady downflows seen in most of the transition region, we may be observing a flow of ionized plasma through a region of changing temperature. For explosive events, the lack of observed proper motions suggests that a temperature change takes place on the time scale of the event that leads to its no longer being visible. As the discussion in Chapter 2 showed, rapid changes in the physical conditions can result in departures from ionization equilibrium.

Since temperature and density diagnostics are based on the assumption that the plasma is in ionization balance, any significant departures could lead to errors in estimates of these basic physical parameters of the emitting plasma. Table 5.3 shows the characteristic ionization and recombination times for the ions C III, C IV, and C V at three temperatures centered on the temperature of peak abundance for C IV. The times were calculated assuming an electron density of 10^{10} cm^{-3}, and the rate coefficients given by Shull and Van Steenberg (1982a, b). For a given ion in the table, $\tau_I(z)$ is the time to ionize to the next highest ion and $\tau_R(z)$ is the time to recombine to the next lowest ion. Thus both processes reduce the relative population of the listed ion.

For a steady downflow, the plasma is probably moving into cooler surroundings, so recombination rates are important. Inspection of the table shows that recombination from C IV to C III is relatively rapid, with a characteristic time of only about 4 s at 10^5 K. Thus C IV ions moving to temperatures characteristic of C III should recombine rapidly. At a downflow velocity of 8 km s^{-1}, the C IV would recombine in a characteristic distance of only 32 km. Recombinations from C V to C IV, however, are much slower, due to the absence of significant dielectronic recombinations at these temperatures. If the downflow begins at a temperature high enough to have C V, C VI, or C VII as the dominant ion, $T \gtrsim 2 \times 10^5$ K, then there can be a significant lag in the creation of C IV by recombination from C V. For a characteristic recombination time of 100 s and a downflow velocity of 8 km s^{-1}, the plasma will travel about 800 km before recombining.

If the downflowing plasma begins with a temperature of 2×10^5 K and travels to a region where the temperature is 10^5 K in less than about 100 s, the C IV number density will be less than the equilibrium value. For an 8 km s^{-1} flow this implies that the temperature gradient must not be greater than about 1.2×10^{-3} K cm^{-1} to assure ionization balance. If this temperature gradient remained constant from 10^4 to 10^6 K, it would imply a transition region thickness of 8000 km, or about 11 arc sec. Observations clearly show that large parts of the transition region are thinner than this.

If the plasma is cooling instead of flowing, the recombination times are again the key parameter. For a cooling plasma, however, the important quantity to compare the recombination times with is the cooling time. Overall, however, the effects are the same whether the plasma flows to a cooler region or cools in place. There will still be an underabundance of C IV ions relative to the equilibrium value if the temperature change is too rapid. Dere et al. (1981) have examined a simple cooling case that may be relevant to the transition region.

For an outflow the plasma is probably moving into hotter surroundings, so ionization rates are important. Inspection of the table shows that the ionization time to produce C IV from C III is less than 5 s at the normal equilibrium temperature of C IV. At a velocity of 108 km s^{-1}, the average peak velocity of an explosive event, this time represents a distance of 540 km. If the outflowing plasma begins at the peak temperature of formation of C III, about 70,000 K, and travels to a region where the temperature is 10^5 K or greater in less than about 5 s, the amount of C IV present will be less than the equilibrium value. For explosive events this implies that the temperature gradient must not be greater than about 5.6×10^{-4} K cm^{-1}. A constant temperature gradient of this value would imply a transition region thickness of more than 17,000 km, much greater than the value typically observed at the limb. If the plasma is being heated without significant motion into hotter regions, then the ionization time must be compared with the characteristic heating time.

Joselyn et al. (1979) have made more detailed calculations similar to the rough calculations outlined here for most ions of interest in the transition region and corona. They solved a simplified form of the set of coupled mass conservation equations for steady flows through a range of temperature gradients. The results provide an estimate of the velocity as a function of ion species, temperature, temperature gradient, and electron density at which ionization equilibrium can no longer be assumed. Their results show, for example, that for an electron density of 10^{10} cm^{-3} and a velocity of 100 km s^{-1} in the direction of increasing temperature, a temperature gradient larger than about 4×10^{-4} K cm^{-1} at 80,000 K will lead to departures from equilibrium in C IV, in agreement with our rough estimate.

Departures from ionization equilibrium can affect determinations of physical parameters in several ways. Consider first temperature diagnostics. The primary temperature diagnostic consists simply of assigning to an emitting region observed in a UV emission line the peak temperature of the $G(T)$ function, where

$$G(T) = \frac{n_{\text{ion}}}{n_{\text{el}}} T^{-1/2} \exp\left(\frac{-h\nu}{kT}\right). \tag{5.22}$$

In a plasma flowing toward lower temperatures, the finite recombination times mean that the ion abundance will represent an environment with a higher temperature than the actual local conditions that produce the line excitation. Thus the actual excitation temperature will be lower than the temperature assigned based on ionization equilibrium. In a plasma flowing to a higher-temperature environment, the finite ionization time will result in a similar lag, leading to an underestimate of the local temperature.

Temperature diagnostics based on a ratio of two emission lines from the same ion should be free of the effects of changes in ion abundances. Unfortunately, as the discussion in Chapter 4 showed, these diagnostics are not readily available. It is interesting, however, that all the useful temperature diagnostics based on lines in the UV spectrum of Al III show temperatures that are below the equilibrium value (Keenan et al., 1986b; Doschek and Feldman, 1987). Since downflows are ubiquitous in the quiet Sun, this is the direction of the discrepancy that would be expected between the actual excitation temperature and the temperature of the peak of the $G(T)$ function based on an equilibrium ion abundance.

Emission measure determinations also will be affected by flow induced departures from ionization equilibrium. As we showed in Chapter 4, the emission measure is usually obtained from the expression for the flux in an emission line. For the two-level case and using the Van Regemorter (1962) approximation, we have

$$F = \frac{2.2 \times 10^{-15}}{4\pi R^2} f A_{el} \int g G(T) n_e^2 \, dV, \tag{5.23}$$

with all the temperature dependence in the $G(T)$ function. For the case of the C IV ion, if the plasma is flowing downward to cooler portions of the atmosphere, the actual temperature that should be used in the $G(T)$ function will be smaller than the equilibrium value and the relative ion abundance n_{ion}/n_{el} will be smaller than the equilibrium value. Thus using the equilibrium values for the temperature and relative ion abundance will result in an emission measure that is smaller than the actual value and is placed at too high a temperature.

This is born out by detailed calculations of the ionization balance in downflows through model atmospheres (e.g., Francis, 1981) and in cooling plasmas (e.g., Dere et al., 1981). Francis (1981) found that for a downflow with a characteristic velocity of 6 km s^{-1} in the region of formation of the C IV resonance lines, the temperature of peak concentration of the C IV ion shifted from $\log T = 5.0$ to $\log T = 4.65$. Moreover, the peak ionic abundance dropped by roughly an order of magnitude. He noted, however, that ions that form at like temperatures in equilibrium tended to continue to form at like temperatures in downflowing plasmas. To first order, ions formed at like temperatures in equilibrium also maintained roughly the same ratio of ionization fractions in a downflow. This is primarily due to the relatively slow recombination times to the Li-like stages of abundant ions such as C, N, and O. Once that recombination takes place, all of the other species with lower degrees of ionization reach ionization balance quickly (e.g., Mariska et al., 1982).

For an outflow, the tendency is in the opposite direction. The emission measure will be placed at too low a temperature and the relative ionic abundances will be larger than the equilibrium values, leading to an overestimate of the emission measure if equilibrium values are used. This is also borne out by ionization balance calculations of upflows through model atmospheres (e.g., Dupree et al., 1979; Borrini and Noci, 1982; Mariska et al., 1982).

Density diagnostics also can be affected by departures from ionization equilibrium. Diagnostics based on pairs of emission lines from the same ion will be affected because all the collisional excitation rates which help determine the level populations contain a factor $T^{-1/2}\exp(-\Delta E/kT)$. Typically these collision rates are evaluated at the temperature of the peak in the line $G(T)$ function or just the temperature of the peak ion abundance. For different temperatures of formation the relationship between the ratio of two density-sensitive lines and the electron density will shift by an amount that depends on the details of the particular atomic system.

For the C III $I(1247\text{ Å})/I(1909\text{ Å})$ ratio there is considerable temperature dependence. A typical quiet-Sun ratio of about 0.03 for these lines implies an electron density of 2×10^{10} cm^{-3} at a temperature of 30,000 K. The same ratio implies an electron density of 3×10^{11} cm^{-3} at a temperature of 56,000 K, and an electron density of about 5×10^9 cm^{-3} at a temperature of 1.26×10^5 K (Cook and Nicolas, 1979). This means that in a downflowing plasma a density estimate based on assuming the equilibrium temperature of formation for the 1247 and 1909 Å lines could easily result in more than an order of magnitude underestimate of the electron density. Raymond and Dupree (1978) have also examined the ef-

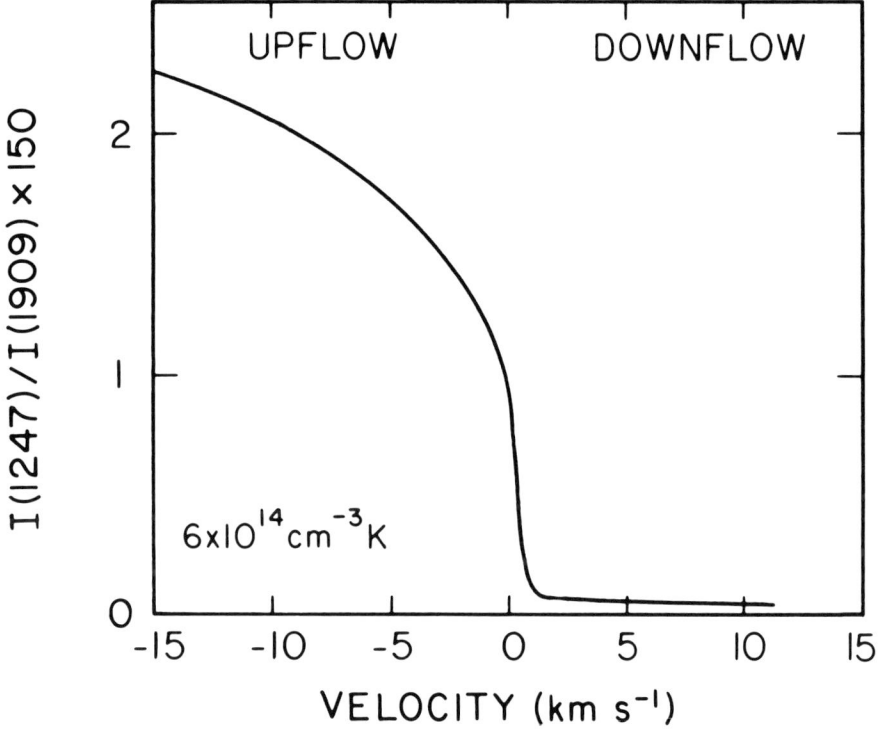

Fig. 5.8. The variation of the C III $I(1247\,\text{Å})/I(1909\,\text{Å})$ ratio for steady mass-conserving flows with the velocities shown through a model transition region with $p/k = 6 \times 10^{14}$ cm^{-3} K (after Raymond and Dupree, 1978).

fects of flows on density diagnostics involving C III. They computed the theoretical line ratio for flows with various velocities through an empirical model atmosphere at constant pressure. Figure 5.8 shows the results of their calculations. For a 10 km s^{-1} upflow, and $n_e T = 6 \times 10^{14}$ cm^{-3} K, they found more than a factor of 2 increase in the observed ratio $I(1247\,\text{Å})/I(1909\,\text{Å})$. For a 10 km s^{-1} downflow, the ratio decreased by more than an order of magnitude. Similarly large changes are present in other density-sensitive ratios involving lines of C III.

Some density-sensitive line ratios are less temperature-sensitive than those involving C III lines. For example the Si III line ratios $I(1301\,\text{Å})/I(1296\,\text{Å})$ and $I(1301\,\text{Å})/I(1303\,\text{Å})$ yield densities that decrease by only 0.1–0.2 dex when the temperature decreases from 50,000 to 32,000 K (Dufton et al., 1983). Other ratios involving lines from the same ion vary by much larger amounts as the temperature changes.

Density diagnostics that involve emission lines from different ions have additional uncertainty. These require knowledge of the level populations of the ion containing the metastable level at the temperature of formation of the intersystem or forbidden line and, since the allowed line typically forms at a different temperature, the shape of the emission measure distribution so as to divide it out. This additional dependence on the emission measure introduces the temperature effects that are important for estimating the emission measure discussed above.

Despite the importance of diagnostics involving emission lines from different ions, there has been little detailed work to assess the effects of flows. Dupree et al. (1979) have calculated the O IV/C IV and O IV/N V ion abundance ratios for an outflow through an empirical quiet-Sun atmospheric model. The mass-conserving outflow had a velocity of 10 km s^{-1} at 10^6 K. Along with the increase of the temperature of formation of each ion, they found that near peak ion abundance the values of the O IV/C IV ratio remain roughly constant, while the O IV/N V ratio shows a decrease of about a factor of 4 relative to equilibrium conditions. For downflows, Francis (1981) performed similar calculations and concluded that, while the temperature of maximum ion abundance and the peak value changed, the ratios of ion abundances for ions formed at nearly the same temperature in equilibrium remained at roughly the equilibrium value. So far, however, there has been no complete study of the effect of flows on density-sensitive line ratios involving different ions. Such a study would require following the time-dependent ionization balance through a variety of flow scenarios and calculating the detailed level populations for the ions involved in the diagnostic ratio.

It might be possible to use a temperature diagnostic coupled with a velocity measurement to estimate the departure from ionization balance and correct the density diagnostic appropriately. Of course for a density-sensitive ratio involving only one ion, a temperature estimate coupled with density sensitivity curves parameterized by temperature should be all that is required. From a practical point of view, the small-scale structuring of the transition region provides yet another complication for any observational analysis and may make unraveling the temperature and density structure from observations alone virtually impossible.

6

Empirical Transition Region Models

So far our examination of the transition region has focused on what we can directly measure or derive from measurements with a minimum of assumptions about the actual physical structure. In this chapter we discuss how to use this information to develop empirical models of the atmospheric structure. In its simplest form, an empirical model is nothing more than an attempt to guess the temperature and density as a function of height in the solar atmosphere. The goal of empirical modelling of the transition region is to take the aggregate of spectroscopic measurements along with the tools that we have developed for extracting physical parameters from it and produce a description of those physical parameters for the observed region. Thus empirical models are descriptive. They contain the fewest assumptions about the underlying physics. Once developed, however, they can provide additional insights into the important physics that may be at work in the transition region.

6.1 Transforming Spectroscopic Data into Empirical Models

Both observations of the transition region and the theoretical tools necessary to understand them are limited. All the UV emission lines formed in the transition region are not simultaneously observable. Different instruments have different spatial, spectral, and temporal resolutions. Their absolute calibrations are often uncertain and change with time. While we understand the basic atomic physics that is at work producing the emission lines, the detailed determination of the necessary atomic parameters is at the forefront of experimental and computational atomic physics. Thus, to make any progress, we must make some simplifying assumptions in applying the tools we are developing for analyzing UV emission-line observations.

First, we assume that the plasma emitting the UV radiation is static. We have already shown that this is not always the case. But most of the observational material we will deal with has such low spectral resolution that it is not possible to measure a Doppler shift. Moreover, without detailed knowledge of the temperature and density structure of the atmosphere, we will be unable to correct for the effects of mass motions. Instead, we assume that flows are not important and then, after we have a clearer picture of what the detailed structure might be, we will examine their effects on our analysis.

Second, in deriving atmospheric parameters, we assume that there are no abundance gradients. Again this assumption is aimed at making an analysis of the observations possible. As we have shown, it is possible to focus an analysis of UV emission lines on accurate abundance determinations and search for variations. To make progress at defining the basic parameters of the transition region, however, we will begin by assuming we know the elemental abundances.

Third, we will assume that no unusual atomic processes are taking place. By unusual, we mean, for example, departures of the electron velocity distribution from a Maxwellian distribution and the effect such a departure can have on collision rates. Charge exchange reactions also fall into this class, though they may also be important. After we have a better idea of the atmospheric structure, we can examine in more detail the possibility that some of these processes are at work.

Finally, we assume that opacity effects can be ignored in analyzing UV emission lines and deriving the atmospheric structure. As we saw earlier, this is not a problem for much of the transition region. Only when we are dealing with strong resonance lines and long path lengths, such as those found in observations at the solar limb, is there a problem.

While emission-line intensities are the raw material for this modelling, they are often used in the form of the derived emission measure. Thus the available data consist of either a set of emission-line intensities or the derived distribution of $\int_R n_e^2 \, dV$ with temperature, along with electron density measurements at one or more temperatures. To make further progress, we must make some assumptions about the geometry of the emitting structures. Early analyses of emission measures (e.g., Pottasch, 1964; Athay, 1966; Dupree and Goldberg, 1967) used whole-disk observations and assumed the geometry was plane parallel. Although we know from the observations presented in Chapter 3 that this may not be a good assumption, we use it here. Since we are using spatially resolved observations, we are assuming that the structures at the spatial resolution of the data can be represented by a plane-parallel model. Thus we have $dV = A \, dh$, where A is the projected area of the spectrograph slit on the Sun and h is the height in the solar atmosphere above some reference height.

To derive the run of temperature and density as a function of height, we also require some relation describing how the density will vary with height. For an atmosphere in hydrostatic equilibrium in a gravitational field, the inward force of gravity balances the outward directed pressure force, leading to

$$\frac{dP}{dh} = -\rho g_\odot, \tag{6.1}$$

where $g_\odot$ is the gravitational acceleration at the surface of the Sun (2.74×10^4 cm s^{-2}). Combining this with the perfect gas law, we have

$$\frac{d\ln P}{dh} = -\frac{\mu m_p g_\odot}{kT}. \tag{6.2}$$

For a constant temperature, this integrates to the barometric equation

$$P = P_0 \exp(-h/H_P), \tag{6.3}$$

where the pressure scale height is

$$H_P = \frac{kT}{\mu m_H g_\odot}. \tag{6.4}$$

For a fully ionized plasma of solar composition, $H_P \approx 5000\,T$ cm. Thus at 10^5 K in the transition region $H_P \approx 5000$ km and higher in the atmosphere at 10^6 K, $H_P \approx 50,000$ km. Since the bulk of the emitting material at the limb falls within this height range, many investigators (e.g., Athay, 1966; Dupree, 1972) have simply assumed that the pressure is constant throughout the transition

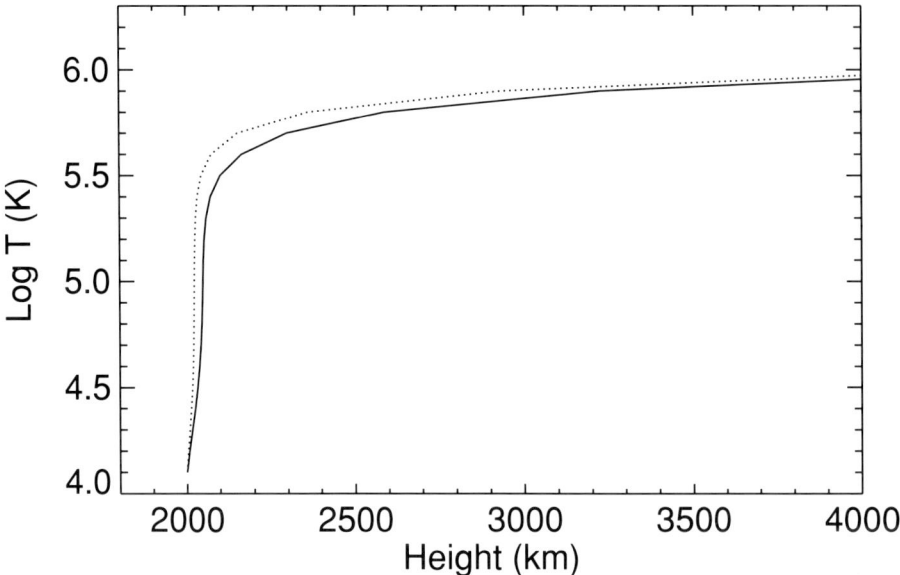

Fig. 6.1. Network (solid line) and cell-center (dotted line) empirical atmospheric models calculated using $\log n_e T = 15.0$ and the emission measures tabulated by Raymond and Doyle (1981b). We have set the height of the $\log T = 4.1$ point to 2000 km.

region. The graph of electron density measurements as a function of temperature shown in Figure 4.11 further supports this approximation.

With the additional assumption of constant pressure, the temperature distribution immediately follows from the emission measure distribution. Introducing the geometry and the assumed stratification of the atmosphere gives for each emission line

$$\int_R n_e^2 \, \mathrm{d}h = \frac{P_0^2}{T^2} \frac{\mathrm{d}h}{\mathrm{d}\log T} \Delta \log T, \tag{6.5}$$

where $P_0 = n_e T$ and the width of the temperature range R has been set to $\Delta \log T$ centered on the temperature of line formation. Each emission line thus provides an estimate of the temperature gradient at the temperature of formation of the line.

Because of the scatter in the individual emission measures, it is generally easier to determine a smooth fit to the individual measurements and then work from that. Figure 6.1 and Tables 6.1 and 6.2 present the results of performing this analysis on the aver-

Table 6.1. *Network empirical model.*

log T (K)	log ∫ n_e^2 dh (cm^{-5})	dT/dh (K cm^{-1})	h (km)	F_c (erg cm^{-2} s^{-1})
4.1	27.60	4.6(−3)	0.0(+0)	9.0(+1)
4.2	27.54	4.2(−3)	7.5(+0)	1.5(+2)
4.3	27.38	4.8(−3)	1.7(+1)	3.0(+2)
4.4	27.13	6.8(−3)	2.6(+1)	7.5(+2)
4.5	26.80	1.2(−2)	3.3(+1)	2.3(+3)
4.6	26.52	1.7(−2)	3.9(+1)	6.1(+3)
4.7	26.14	3.3(−2)	4.3(+1)	2.1(+4)
4.8	25.75	6.5(−2)	4.6(+1)	7.1(+4)
4.9	25.42	1.1(−1)	4.8(+1)	2.2(+5)
5.0	25.03	2.1(−1)	4.9(+1)	7.5(+5)
5.1	25.06	1.6(−1)	5.1(+1)	9.9(+5)
5.2	25.17	9.8(−2)	5.4(+1)	1.1(+6)
5.3	25.31	5.7(−2)	6.0(+1)	1.1(+6)
5.4	25.45	3.3(−2)	7.2(+1)	1.1(+6)
5.5	25.60	1.8(−2)	1.0(+2)	1.1(+6)
5.6	25.73	1.1(−2)	1.6(+2)	1.2(+6)
5.7	25.86	6.3(−3)	3.0(+2)	1.2(+6)
5.8	26.00	3.6(−3)	5.9(+2)	1.3(+6)
5.9	26.14	2.1(−3)	1.2(+3)	1.3(+6)
6.0	26.29	1.2(−3)	2.6(+3)	1.3(+6)
6.1	26.43	6.8(−4)	5.7(+3)	1.3(+6)
6.2	26.48	4.8(−4)	1.2(+4)	1.7(+6)
6.3	26.42	4.4(−4)	2.1(+4)	2.7(+6)

age quiet-Sun cell-center and network emission measures derived by Raymond and Doyle (1981b) and shown in Figure 4.1. Based on the measured electron density of 1.8×10^{10} cm^{-3} at 60,000 K, we have calculated these models assuming a value for P_0 of 10^{15} cm^{-3} K. In addition, since the Raymond and Doyle (1981b) computed emission measures for a logarithmic temperature interval of 0.1 dex, the value of $\Delta \log T$ is set to 0.1, instead of the more common value of 0.3 used in most emission measure calculations.

These models have all the characteristics of other empirical models based on low-resolution UV observations. There is a rapid rise

Table 6.2. *Cell-center empirical model.*

log T (K)	log ∫ n_e^2 dh (cm^{-5})	dT/dh (K cm^{-1})	h (km)	F_c (erg cm^{-2} s^{-1})
4.1	27.34	8.4(−3)	0.0(+0)	1.6(+2)
4.2	27.27	7.8(−3)	4.1(+0)	2.7(+2)
4.3	27.08	9.6(−3)	8.8(+0)	5.9(+2)
4.4	26.84	1.3(−2)	1.3(+1)	1.5(+3)
4.5	26.52	2.2(−2)	1.7(+1)	4.3(+3)
4.6	26.05	5.2(−2)	2.0(+1)	1.8(+4)
4.7	25.64	1.1(−1)	2.1(+1)	6.5(+4)
4.8	25.32	1.7(−1)	2.2(+1)	1.9(+5)
4.9	25.03	2.7(−1)	2.3(+1)	5.3(+5)
5.0	24.84	3.3(−1)	2.4(+1)	1.2(+6)
5.1	24.80	2.9(−1)	2.4(+1)	1.8(+6)
5.2	24.81	2.3(−1)	2.6(+1)	2.5(+6)
5.3	24.92	1.4(−1)	2.8(+1)	2.7(+6)
5.4	25.02	8.8(−2)	3.3(+1)	3.0(+6)
5.5	25.24	4.2(−2)	4.5(+1)	2.6(+6)
5.6	25.43	2.1(−2)	7.5(+1)	2.4(+6)
5.7	25.64	1.1(−2)	1.5(+2)	2.1(+6)
5.8	25.89	4.7(−3)	3.6(+2)	1.6(+6)
5.9	26.12	2.2(−3)	9.3(+2)	1.4(+6)
6.0	26.32	1.1(−3)	2.4(+3)	1.2(+6)
6.1	26.43	6.8(−4)	5.6(+3)	1.3(+6)
6.2	26.45	5.2(−4)	1.1(+4)	1.8(+6)
6.3	26.39	4.7(−4)	2.0(+4)	2.9(+6)

in the temperature through the transition region, with the peak temperature gradient occurring at 10^5 K. As the temperature approaches coronal values, the temperature gradient becomes much smaller. In both models the thickness of the transition region is small. Taking the lower boundary as 25,000 K and the upper boundary as 10^6 K, the thickness is only 2600 km for the network model and 2400 km for the cell-center model. As the figure vividly shows, however, the jump from 25,000 to 3×10^5 K takes place in less than 100 km.

Also included in the tabulations is the value of the thermal conductive flux F_c at each location, where
$$F_c = \kappa T^{5/2} \frac{dT}{dh}, \qquad (6.6)$$
with $\kappa = 1.1 \times 10^{-6}$ (Ulmschneider, 1970). This represents the amount of thermal energy being carried from the corona through the transition region. Well before the advent of solar UV emission-line data, Giovanelli (1949) argued that the transition region structure should be dominated by thermal conduction and that F_c should be constant. Simple empirical models like the ones derived above support this suggestion. For the network model the conductive flux is nearly constant at a value near 1.1×10^6 erg cm^{-2} s^{-1} from $\log T = 5.2$ to close to coronal temperatures. For the cell-center model the conductive flux is closer to 2×10^6 erg cm^{-2} s^{-1} in this range and varies more.

Athay (1966) first noted observationally this near constancy of the conductive flux. He assumed that the pressure was constant over the entire transition region and that the temperature gradient was constant over the region of line formation and rewrote the integral in the equation for the emission from a two-level atom as
$$\int_R gG(T) n_e^2 \, dh = P_0^2 \left\langle \frac{dT}{dh} \right\rangle^{-1} \int_R gG(T) T^{-2} \, dT. \qquad (6.7)$$
Plotting $(dT/dh)^{-1}$ against temperature, Athay noticed that in the 10^5–10^6 K range the temperature gradient followed the relationship
$$T^{5/2} \frac{dT}{dh} = \text{constant}. \qquad (6.8)$$
If there is no magnetic field present, this relationship is the condition for a constant thermal conductive energy flux.

Dupree (1972) incorporated this near constancy of the conductive flux into her analysis of *OSO 4* spectra. She rewrote the integral in the emission measure equation as
$$\int_R gG(T) n_e^2 \, dh = \frac{P_0^2}{F_0} \int_R g \frac{n_{\text{ion}}}{n_{\text{el}}} \exp\left(\frac{-h\nu}{kT}\right) dT, \qquad (6.9)$$
where
$$F_c = \kappa F_0 = \kappa T^{5/2} \frac{dT}{dh} \qquad (6.10)$$
is assumed to be constant over the region of line formation and the extra factor of $T^{-1/2}$ comes from the $G(T)$ function. In the

temperature range from 10^5 to 10^6 K, all the data points calculated in this manner formed a horizontal band in a plot of P_0^2/F_0 against temperature, again showing the near constancy of the conductive flux.

Explicitly removing P_0^2/F_0 from the integral highlights the fact that the two quantities are closely related. With the assumption that P_0 is constant throughout the transition region, the slope of the emission measure curve indicates whether the conductive energy flux is constant. In Dupree's (1972) analysis, a horizontal distribution of points in a plot of P_0^2/F_0 against T shows that F_0 is constant. In the analysis discussed earlier, which is similar to the formulation used by Jordan and Wilson (1971), a slope of 1.5 in a plot of $\log \int_R n_e^2 \, dh$ against $\log T$ indicates constant conductive flux.

Actually determining the value of F_c, however, requires an estimate of the pressure. Since we need the square of the pressure, small uncertainties in the magnitude of the pressure are magnified. The estimate of $\log P_0 = 15.0$ used in generating the models in Tables 6.1 and 6.2 produces conductive fluxes in excess of 10^6 erg cm^{-2} s^{-1}. A lower estimate of $\log P_0 = 14.8$, the value at the top of the average network model of Vernazza et al. (1981), yields a conductive flux in the network of about 4.5×10^5 erg cm^{-2} s^{-1}. Using the Vernazza et al. estimate of $\log P_0 = 14.6$ for the cell interior yields a conductive flux in the cell interior model of about 3×10^5–4×10^5 erg cm^{-2} s^{-1}. Thus the typical factor-of-2 errors that are often attributed to transition region density determinations can result in a factor-of-4 error variation in estimates of the conductive energy flux.

There are many possible refinements to the process of constructing a model atmosphere from the data. For example, rather than assume a constant pressure throughout the transition region, we could assume that the pressure is constant over just the region of formation for each emission line. Given a starting pressure, successive iterations between the emission measure equation and the equation of hydrostatic equilibrium then result in the run of the temperature gradient and pressure gradient with height (e.g., Jordan and Wilson, 1971). A second refinement is to use the derived model atmosphere to compute the emission in each emission line used in the emission measure analysis. The model can then be adjusted to obtain an improved fit to the observations (e.g., Raymond and Foukal, 1982). This helps account for the fact that the emission measure for each line is computed at a single temperature, but the line actually forms over a range of temperatures with varying densities. Emission lines from Li-like ions are particularly prone to this problem, as are some intercombination lines.

6.2 Effect of Inhomogeneous Structures

Since the models we have just derived are based on average low-resolution observations, they do not incorporate any of the fine-scale structure or dynamics observed at higher spectral or spatial resolution. If, however, the structuring and dynamics we observe are perturbations imposed on a simple background atmosphere, then the models would provide a useful first step for understanding the physics of these layers of the atmosphere. Here we examine attempts to estimate the contributions from small-scale structures.

6.2.1 Contribution of Spicule-Like Structures

On the disk even at moderate spatial resolution, it is difficult to learn much about the structures present within the network at transition region temperatures. At the limb, however, much of this structuring becomes evident. Thus comparison of the simple models outlined above with limb observations provides one way of gauging the impact of inhomogeneities on these models.

One method for comparing the models with observations at the limb is to compute theoretical limb brightening curves from the models and then compare those curves with observed average limb brightening curves. Assuming that the model atmosphere is spherically symmetric, the flux in an optically-thin line at any position on or above the solar disk is given by

$$F(\rho) = 2.2 \times 10^{-15} \frac{A}{4\pi R^2} A_{\text{el}} f \int_{R_0}^{\infty} gG(T) n_e^2 \frac{dr}{\mu}. \qquad (6.11)$$

Here ρ is the distance from Sun center in units of the solar radius, r is the radial distance, A is the area of the slit at the Sun, R_0 is the height in the model corresponding to the line of sight at height ρ, and

$$\mu = \left[r^2 - (\rho R_\odot)^2 \right]^{1/2} / r \qquad (6.12)$$

if $\rho < 1$ and one half that quantity if $\rho \geq 1$. This change above the limb accounts for the doubling of the path length when the solar surface no longer blocks the emission from behind the Sun. Inside the limb, the lower limit to the integrand is the base of the model atmosphere. Outside the limb, however, it moves upward in the model atmosphere as the value of ρ increases.

Mariska and Withbroe (1975) applied this technique to 5 arc sec resolution EUV data from the Harvard experiment on *Skylab*. They found that an average atmospheric model with $\log P_0 = 14.8$, a coronal temperature of 1.1×10^6 K, and a constant conductive flux of 3.5×10^5 erg cm^{-2} s^{-1} provided the best fit to average quiet-Sun limb brightening data in the O VI 1032 Å and Mg X 625 Å resonance lines. To account for the disk-center differences in network and cell-center emission in the O VI resonance line, they suggested a network model with a constant conductive flux of 3.5×10^5 erg cm^{-2} s^{-1} and a cell-center model with a constant conductive flux of 1.1×10^6 erg cm^{-2} s^{-1}. These values are near those listed in Tables 6.1 and 6.2, when the lower pressure is accounted for. The solid line in Figure 6.2 shows their fit to the O VI data. This two-component model still, however, failed to account for the emission observed in the O VI 1032 Å emission line immediately above the limb, where more emission was observed than was predicted in the model. Mariska and Withbroe (1975) suggested that the excess emission was due to spicule-like inhomogeneities.

Withbroe and Mariska (1976) found that a model with a spicule component contributing 20% of the total quiet-Sun emission for EUV lines of the C II, C III, N III, and O VI ions could account for all the emission. These EUV-emitting spicules had the same height variation as Hα-emitting spicules. This model is shown as the dashed line in Figure 6.2.

At higher spatial resolution these simple two-component models fail to match the observations. Mariska *et al.* (1978) applied this model to 2 arc sec resolution NRL *Skylab* limb brightening data. They found that neither a spicule model nor a two-component model would fit the observations. The scale height for the decrease in emission above the limb clearly suggests, however, that spicule-like emitting structures play some role. Kanno (1978) also applied the Withbroe and Mariska (1976) spicule model to NRL *Skylab* data. He found that for temperatures between 30,000 and

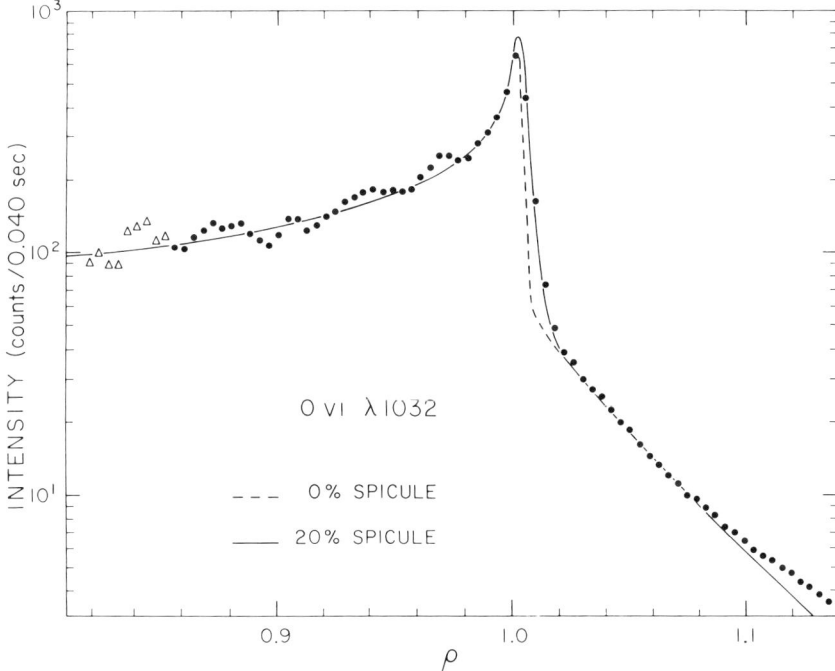

Fig. 6.2. Comparison of limb brightening measurements for the O VI 1032 Å line (points) with a theoretical plane-parallel model and a plane-parallel model combined with a 20% contribution from spicule-like structures (from Withbroe and Mariska, 1976).

2×10^5 K, nearly all the emission was from a spicule-like component. At higher and lower temperatures, the fraction decreased toward the 20% value found by Withbroe and Mariska (1976). Uncertainties in the data are large enough, however, that the main conclusion we can draw is that the *Skylab* limb data are not consistent with either a simple spherically symmetric model or with a spicule model.

Feldman *et al.* (1979) attempted to resolve the question of the role of inhomogeneous structures in the transition region by inverting the observed emission above the limb to derive a source function for the emitting plasma. They found that the spicule-like component, which they assumed was responsible for nearly all the emission, occupies only about 1% of the solar surface at a temperature of 10^5 K. This suggests that the assumption that we can model the atmosphere as a simple homogeneous, radially stratified structure is seriously flawed. The models derived in Tables 6.1 and 6.2 are clearly compromised by a lack of detailed geometrical information.

As the spatial resolution of the observations has improved, this question of the detailed geometry of the emitting layers has become more important. As we already discussed in Chapter 3, high spatial resolution observations in the C IV resonance lines show that a typical discrete structure is about 2400 km across (Dere et al., 1987). Assuming that the measured emission from such a discrete structure is from a single layer of thickness Δh, the emission measure in the C IV resonance line becomes

$$\int_R n_e^2 \, dh = \langle n_e^2 \rangle \Delta h. \tag{6.13}$$

Using HRTS observations from *Spacelab 2*, Dere et al. (1987) found electron densities from O IV line intensities that implied scaled electron pressures of 10^{15}–10^{16} cm^{-3} K and path lengths ranging from 0.1 to 10 km. These path lengths agree with those in the models in Tables 6.1 and 6.2, but clearly disagree with the extended emission seen in the C IV resonance lines above the limb.

Dere et al. (1987) suggested that instead of a single thin layer, the material emitting in the C IV resonance lines was arranged as many filamentary structures distributed throughout an Hα emitting spicule. Figure 6.3 shows their general picture. This geometry is based primarily on the observation that the quiet transition region is organized into structures that are extensions of Hα spicules (Dere et al., 1986). If a single UV-emitting spicule contained 10 or more subresolution structures, filling factors of 1% or less would result.

6.2.2 Filling Factors

The whole concept of a filling factor is elusive. Athay (1966) noted that a lack of detailed knowledge of the geometry could add uncertainty to any models constructed using an emission measure analysis. He was working with full disk fluxes and pointed out that if emitting material covered only a fraction of the disk, then the expression for the emission measure should be rewritten as

$$\int_R n_e^2 \frac{dh}{dT} \, dT = \int \gamma n_e^2 \frac{dh}{dT} \, dT, \tag{6.14}$$

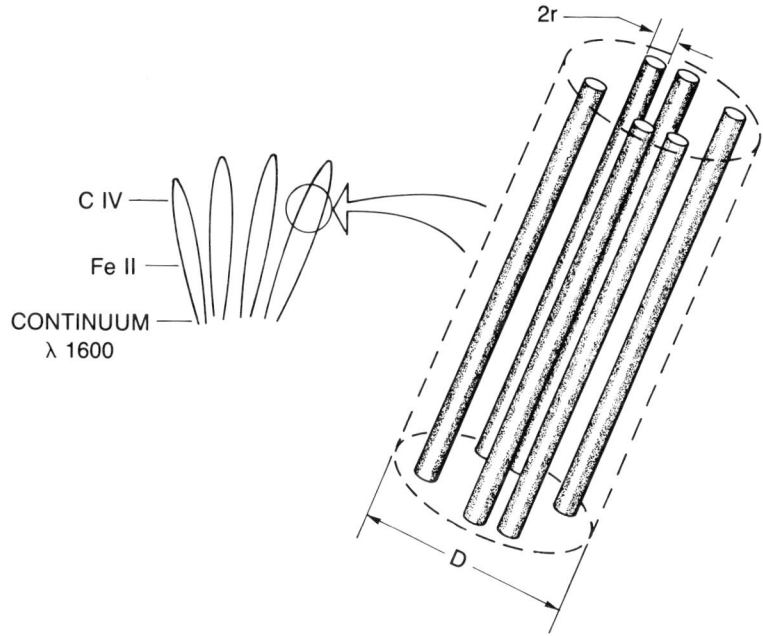

SUBRESOLUTION FILAMENTARY STRUCTURE OF SPICULES AT 10^5 K

Fig. 6.3. One possible model for the small-scale structuring present in the transition region (from Dere et al., 1987).

where γ is the fraction of the solar disk covered with emitting material. Because he lacked transition region density diagnostics and reasonable spatial resolution, Athay was unable to determine whether these inhomogeneities were in the density, the temperature gradient, or both.

More recent observations have the advantage of high spatial resolution and diagnostic lines for obtaining electron densities. Thus the filling factor becomes tied up again in the question of the geometry of the emitting plasma. Feldman et al. (1979) and Dere et al. (1984) compared observations at the limb with observations at Sun center to derive an areal filling factor. The emission in one line at the limb provides an estimate of the radial distribution of material at the temperature of formation of the line as viewed tangentially with a spectrograph aperture of some fixed area. The emission at Sun center provides an estimate of the emission from the same temperature range as viewed radially. If the emitting material is distributed in a spherically-symmetric shell, then the model derived from the limb observations should predict the intensities observed at Sun center. In fact the computed Sun-center intensities are too large. They can be brought into agreement with the limb-based model by assuming that only some fraction γ of the volume represented by the spherical shell with a thickness equal to the radial extent of the emitting region at the limb is actually occupied by emitting plasma.

Feldman et al. (1979) obtained values for γ of about 0.01 for 10^5 K plasma at a height of 2–4 arc sec above the limb. They derived this result from observations made with a spectrograph slit that was 2 arc sec × 60 arc sec in size. Dere et al. (1984) made the same limb to disk comparison using data with 1 arc sec × 1 arc sec resolution and obtained a filling factor of 0.16. Both estimates assume that spicule-like structures produce the emission. In addition, they assume that most of the emission produced at the limb is from structures originating in the transition region network, and hence use disk-center network intensities in the calculation. Thus they do not consider any homogeneous spherically-symmetric component that might be present in the cell interiors and possibly also in the network.

Using cell-center line intensities, Feldman et al. (1979) derived an approximate thickness for this component of only 1 km at the temperature of formation of C III. This is not wildly different from the thickness of the C III-emitting region in the empirical cell-center model in Table 6.2. Their final model then had a homogeneous thin component that accounts for a small fraction of the total emission and a spicule-like component occupying 1% of the surface area. Thus every line of sight to a point on the disk near Sun center would pass through transition region temperature material, but only 1% of the time would that line of sight intersect one of the structures responsible for the bulk of the emission. Near the limb more of the spectrograph aperture would be filled with the structures responsible for the bulk of the emission because of projection effects.

The concept of a filling factor also has been applied to high spatial resolution observations of individual features on the disk. Dere et al. (1987) for example assumed that although a typical emitting element on the disk is about 2400 km in length along the HRTS slit,

Effect of Inhomogeneous Structures

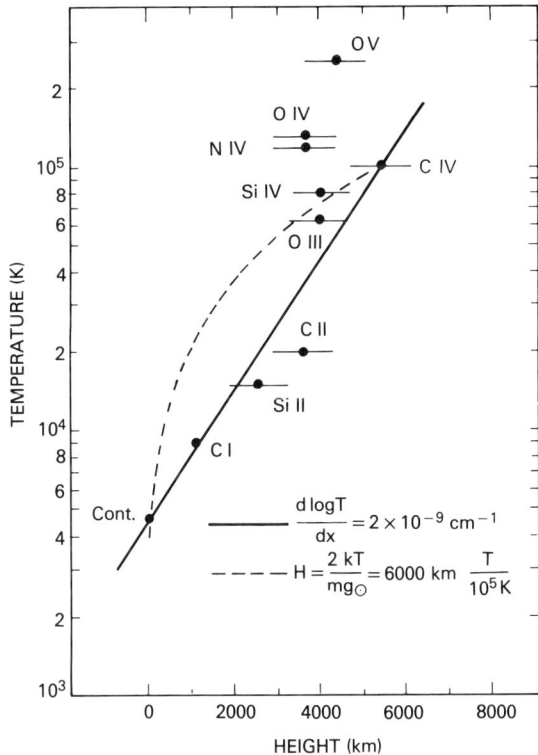

Fig. 6.4. Measurements near the limb of the height separation of transition region emission lines (from Cook and Brueckner, 1991).

the emitting regions within that element are actually filamentary structures. They then derived areal filling factors of 10^{-5}–10^{-2}. This means that a spectrograph with sufficiently high spatial resolution would see the 2400 km structure break up into filaments, each of which had a radius of only 3–30 km. Such small filling factors, even with 1 arc sec resolution data, suggest that it will be very difficult to deduce the geometry of the structures that make up the transition region from spatial observations alone.

On the other hand, there are observations near the limb that show height differentiation between lines formed at different temperatures in the transition region. Figure 6.4 shows one case from the HRTS data (Cook and Brueckner, 1991). Here there is a clear separation in height between C I emission lines and lines formed at higher temperatures from ions ranging from C II to O V. While there is a separation between the Si II and C II emission lines, the data from the hotter lines are consistent with their being formed over a very narrow height range. Dere *et al.* (1986) also found a distinct separation between the location of the emission from cool chromospheric lines and the location of emission from the C IV resonance lines. They made no measurements, however, of separations between transition region lines. Observations made above the limb, such as those shown in Figure 3.5, also show height separations between lines formed at different temperatures.

While the concept of a filling factor is a useful one, we must bear in mind that it always involves some picture of the underlying geometry. This assumed geometry determines to some extent the derived value of the filling factor. Ultimately of course the goal is to obtain observations at sufficiently high spatial resolution that the filling factor of each observed structure is unity and the true geometry of the transition region becomes apparent.

6.2.3 Absorption by Inhomogeneous Structures

Near the limb spicule-like inhomogeneities can affect the observed emission in two ways. For wavelengths longward of the Lyman limit at 912 Å, the increased emission they produce at the limb results in the extended emission we see just above the limb. For wavelengths shortward of the Lyman limit, they not only contribute this extended emission, but also absorb radiation through opacity in the Lyman continuum. Thus we would expect that the limb brightening curves for emission lines formed at the same temperature but above and below the Lyman limit should be different, with the difference showing the amount of neutral hydrogen in the spicule-like component (Withbroe, 1970*b*; Withbroe and Mariska, 1976).

Withbroe and Mariska (1976) saw a sizable difference in the limb brightening curves between the N III 991 Å emission line and the O IV 554 Å emission line, which they attributed to absorption by the cool cores of spicule-like structures. They were, however, unable to reconcile fully the decrease in emission at the limb with the results obtained from lines formed longward of the Lyman limit. While the long wavelength results suggested that spicule-like structures produced 20% of the total emission, the O IV data required a 60%

or greater spicule contribution. This again suggests that a spicule model for the EUV-emitting structures may be too simple.

Using an extension of the same model, Kanno (1983) was unable to match the behavior of the ratio of the O IV 554 Å line to the O VI 1032 Å line near the limb. The data suggested that a distribution of absorbers at 5 arc sec (3600 km) above the white-light limb would provide a better description than a spicule-like distribution. To a first approximation this ratio should be sensitive to the opacity in the Lyman continuum. Since the two lines form at different temperatures and the O VI line has a significant contribution from coronal temperature plasma, the result must be viewed with some caution. There is no doubt, however, that the Harvard *Skylab* observations of limb brightening differ for similar emission lines formed above and below the Lyman limit, suggesting that absorption by neutral hydrogen is present somewhere in the height range over which the EUV lines form.

Lyman continuum absorption also may be present on the disk near Sun center. Kanno (1979) and Schmahl and Orrall (1979) first reported evidence for this and estimated Sun-center opacities at 912 Å of 2 and 4.6, respectively. Doschek and Feldman (1982) on the other hand examined a combined set of data from both the Harvard and NRL experiments on *Skylab* using improved atomic data and found no wavelength dependence for the absorption. Since the absorption cross-section is proportional to λ^3, they argued that the apparent absorption reported by Kanno and Schmahl and Orrall was primarily due to their use of less accurate atomic data along with uncertainties in the instrumental calibrations. Raymond and Doyle (1981*b*) also concluded that the absorption was negligible. They suggested that the apparent absorption seen in the C II lines used by Kanno (1979) was the result of their being formed at a lower temperature than indicated by the empirical models.

The origin of the discrepancies in line ratios between emission lines formed at wavelengths above and below the Lyman limit is still not fully resolved. Kanno and collaborators (Kanno and Suematsu, 1982; Kanno *et al.*, 1984; Nishikawa, 1983) have extended the earlier work and argued that the effect is real and the result of clouds of absorbing neutral hydrogen instead of spicules. Their current estimate of the quiet-Sun optical thickness of these clouds at 912 Å at disk center is about 1.6 (Kanno *et al.*, 1984). The lack

182 *Empirical Transition Region Models*

of a clear wavelength dependence, even in the data presented by Kanno, suggests that the reduced emission at wavelengths below 912 Å may be due only in part to neutral hydrogen absorption.

6.3 Consequences of the Empirical Models

While inhomogeneities are clearly an important factor in determining the structure of the transition region, there may still be a component to the emission that satisfies the assumptions that we used to derive the empirical models. If that component is a small fraction of the total emission, then the empirical models will seriously mislead us. If that component is a large fraction, then we need to understand the consequences of these models for diagnosing the transition region plasma. We have already discussed the effect of flows through a steep temperature gradient on the ionization balance and potentially on some temperature diagnostics. Even if the plasma is not flowing, however, the steep temperature gradient inferred from the observations can result in other deviations from the assumptions used in the analysis. Of primary concern are departures from a Maxwellian electron energy distribution and diffusion. These effects can potentially change the atomic ionization and excitation equilibria, resulting in errors in the derived emission measures upon which the empirical models are based.

6.3.1 Non-Maxwellian Electron Energy Distributions

Non-Maxwellian electron energy distributions in the transition region result from hot coronal electrons diffusing downward into the transition region. There they produce a high-velocity tail in the local electron energy distribution. These energetic electrons produce heat fluxes that depart from classical values, which are calculated by assuming a collision-dominated plasma, and alter ionization, recombination, and excitation rates, which are computed assuming a Maxwellian distribution function.

The mean free path of a thermal electron λ_e is simply the thermal velocity v_{th} multiplied by the mean time between collisions τ. From Braginskii (1965) we have

$$\lambda_e \approx \frac{1.1 \times 10^5 T^2}{n_e \ln \Lambda} \quad \text{cm}, \tag{6.15}$$

where T is in K. Using the scaled electron pressure to eliminate the electron density, we have

$$\lambda_e \approx \frac{1.1 \times 10^5 T^3}{P_0 \ln \Lambda}. \qquad (6.16)$$

Thus for constant pressure transition region models such as those in Tables 6.1 and 6.2, λ_e is proportional to T^3. At 10^5 K, $\lambda_e = 5.5 \times 10^3$ cm, while at 10^6 K it has increased to 5.5×10^6 cm.

More important, however, is the ratio of the mean free path to the temperature scale height in the corona,

$$\alpha = \lambda_e \frac{d \ln T}{dh}. \qquad (6.17)$$

For the models in Tables 6.1 and 6.2, this ratio increases from $\sim 10^{-5}$ at low temperatures to a maximum of $\sim 10^{-2}$ between 10^5 and 3×10^5 K where the temperature gradient is steepest, and then decreases again to $\sim 5 \times 10^{-3}$ at the corona. Thus it would appear that although there are large temperature gradients in the transition region, the mean free paths are small enough that the electron distribution function should remain Maxwellian. Unfortunately, for values of $\alpha \gtrsim 10^{-3}$ this assumption begins to break down (e.g., Shoub, 1983).

There have been several attempts to evaluate the consequences of this breakdown (Gurevich and Istomin, 1979; Roussel-Dupré, 1980b, c; Owocki and Scudder, 1983; Shoub, 1983). Roussel-Dupré (1980b, c) and Shoub (1983), using different techniques, have calculated electron velocity distributions in simple static transition region models. Roussel-Dupré (1980b), using a velocity distribution consisting of a thermal part and a collisionless part, found enhancements of the heat flux in the transition region, but suggested that the effect was small. Shoub (1983) performed a more detailed calculation and found that the enhanced heat flux was significant.

Both Roussel-Dupré (1980c) and Shoub (1983) found that the high-energy electrons could result in significant departures from ionization equilibrium of trace elements throughout the transition region. Keenan (1984) applied Shoub's results to the calculation of collision rates for the Be-sequence ions C III and O V. He found no significant changes for the plasma conditions where the ions are most abundant, but enhanced rates at lower temperatures where the ion abundances are small. Dufton et al. (1984) found, however, that inclusion of Shoub's results in the calculation of collision rates for Si III removed a previous discrepancy between atomic physics calculations and solar UV observations. Thus for simple static empirical models, energetic electrons from the corona can have some effect on both the heat flux and the ionization and excitation equilibria. Full understanding of the detailed consequences must, however, await further theoretical work.

6.3.2 Diffusion

Whenever there is a gradient present in a physical quantity describing a neutral gas or a plasma, there is the possibility of the transport of some physical property of the medium in an effort to reduce that gradient. Ordinary diffusion is the transport of a species from a region of high concentration to a region of low concentration because of a gradient in the species concentration. Thermal diffusion is the transport of a species because of the presence of a temperature gradient. In a plasma the electrical properties of the medium alter the rate of diffusion. For example near the base of the transition region there is a substantial population of neutral atoms. In this situation the more mobile electrons will diffuse faster than the heavier ions. This leads to charge separation, which produces an electric field that acts to retard the diffusion of the electrons and enhance the diffusion of the ions, resulting in charge neutrality. This enhanced ion diffusion is referred to as ambipolar diffusion. Here we consider studies of diffusion in empirical atmospheric models. In the next chapter we will also consider studies of the role of diffusion in determining the structure of physical transition region models.

There have been several studies of diffusion in both static atmospheres and atmospheres with flows. Tworkowski (1975, 1980) found that diffusion tended to move ions to temperatures higher than their equilibrium values and could result in significant variations in elemental abundances. Using new calculations of diffusion coefficients for ions of helium, carbon, nitrogen, and oxygen (Roussel-Dupré, 1981), Roussel-Dupré (1980a), and Roussel-Dupré and Beerman (1981) found essentially the same result. They examined the effects of various mass conserving flows acting with diffusion and found that the flows often dominated over diffusion.

For the case of no flow, Roussel-Dupré and Beerman (1981) calculated depletions of almost two orders of magnitude in the carbon and silicon abundances in portions of the transition region, with corresponding increases in the corona. For small flow velocities, however, the deviations in abundance were a factor of 3 or less, with the downflowing model enhancing the transition region abundances.

Besides the change in elemental abundances, any flow changes the ionization balance. Here the effects are similar to those we have already noted. Downflows tend to reduce the temperature of peak ion concentration, while upflows have the opposite effect.

Because diffusion can produce gradients in elemental abundances, it may be a factor in the abundance variations observed in the solar atmosphere. Diffusion is dependent on ionic charge, but there is no dependence on first ionization potential. Since the basic variation observed is a coronal depletion of heavy elements with first ionization potentials greater than about 9 eV, more than diffusion must be operating.

Like studies of non-Maxwellian energy distributions, studies of the role of diffusion in empirical atmospheric models are just beginning. Both effects may be important. How important they are depends on how much of the solar atmosphere, if any, conforms to simple empirical models such as the ones we have derived here.

6.4 Empirical Energy Balance

As we pointed out in Chapter 1, a major goal of solar physics is to understand the flow of mass and energy from beneath the chromosphere to the corona and solar wind. Because we assumed in our analysis that the transition region plasma is static, the energy balance is quite simple. Only three processes are important in the static energy balance: conduction, radiation, and heating.

For an optically-thin, static, transition region plasma, the equation for the conservation of energy at each location is

$$\nabla \cdot \mathbf{F}_c + E_h + E_r = 0, \tag{6.18}$$

where $\mathbf{F}_c$ is the thermal conductive flux, E_h is the local energy deposition rate due to the heating mechanism, and E_r is the local energy loss rate due to radiation. We assumed in our empirical model that the atmosphere is one-dimensional, so the divergence of the conductive flux becomes

$$\frac{\mathrm{d}F_c}{\mathrm{d}h} = \frac{\mathrm{d}}{\mathrm{d}h}\left(\kappa \frac{\mathrm{d}T}{\mathrm{d}h}\right). \tag{6.19}$$

Since one result of the emission measure analysis is the temperature gradient, we know all the information necessary to compute the thermal conductive flux and its divergence.

The local radiative loss rate at each location follows immediately by using the data in Table 2.4 and the temperature and density at each location in the models. Finally, combining those two rates results in the local energy deposition requirements for the empirical models.

6.4.1 Global Energy Balance

Analyzing the local energy balance of an empirical model is a hazardous undertaking. We have already ignored scatter in the data points in producing a smooth emission measure distribution, and then the slope of that distribution is an important determinant of the energy balance. Despite these uncertainties, if we consider the transition region energy balance implied by the models as a whole, there is still some useful information that we can use as a starting point for more complex theoretical energy balance models.

At around 10^6 K, the models in Tables 6.1 and 6.2 show a downward thermal conductive flux of about 10^6 erg cm^{-2} s^{-1}. Because of the much shallower temperature gradients at the base of the transition region and the strong temperature dependence of the conductivity, the downward conductive flux at the base of the transition region is essentially zero. Moreover, at least at the spatial resolution currently available in the 10^6 K temperature range, the observations suggest that this downward energy flux applies uniformly in space. Thus the models show that without considering any additional heating source, there is an energy input to the transition region at its top of about 10^6 erg cm^{-2} s^{-1}, and a corresponding energy loss from the corona.

High-resolution observations at 10^6 K may very well show that the small filling factors observed in the transition region persist into the corona. The local downward conductive energy flux would then remain near 10^6 erg cm^{-2} s^{-1}, but the global energy lost from the corona and available to the transition region would be reduced considerably. For example, if the apparent transition region filling factor of 1% persists into the corona, the globally averaged energy lost from the corona and available to the transition region would be only 10^4 erg cm^{-2} s^{-1}. This of course leaves open the issue of the nature of the plasma between the inhomogeneous structures and how it contributes to the energy balance. Because we have no high-resolution coronal observations to guide us, it is equally possible that at coronal temperatures we are seeing regions that, because of the magnetic field geometry, are not physically connected to the small-scale structures that may dominate the transition region emission. In that case the larger conductive energy flux estimate would be more appropriate. Without a firmer understanding of the geometry, all we can do is place limits on the conductive flux from the corona.

If we ignore the filling factor issue and simply add up the radiative losses in the models, we immediately have some idea of the heating requirements for these layers. Integrating from the 10^6 K level down to the base of the model, the total radiative flux is 1.9×10^6 erg cm^{-2} s^{-1} in the network model, and 9.9×10^5 erg cm^{-2} s^{-1} in the cell-center model. Both numbers are remarkably close to the conductive flux entering from the corona, suggesting that the energy balance in the transition region is quite simple; energy enters the transition region by conduction from the overlying corona and leaves as radiation by the time it reaches the base of the transition region. Whatever is heating the corona is passing through the transition region without directly depositing any energy. Unfortunately, this picture does not hold up in detail when we examine the local energy balance implied by the empirical models.

6.4.2 Local Energy Balance

Figures 6.5 and 6.6 show the derived local energy balance for the network and cell-center models tabulated in Tables 6.1 and 6.2. In these figures, positive values represent cooling, while negative values represent heating. Thus radiation is always positive, while the divergence of the conductive flux can be either positive or negative.

The network and cell models exhibit different behavior in both the corona and the transition region. In the network empirical

188   Empirical Transition Region Models

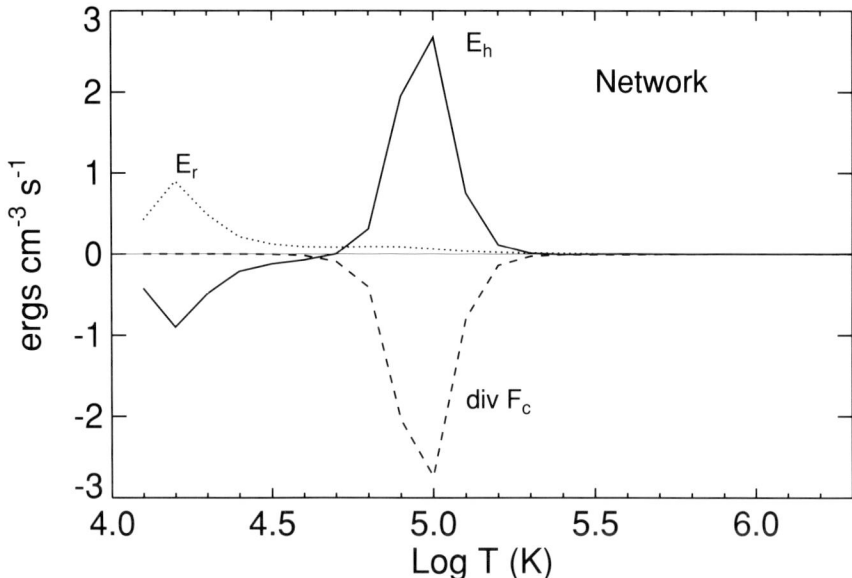

Fig. 6.5. Local network energy balance calculated using the empirical model in Table 6.1. Positive values represent energy losses.

model, the divergence of the conductive flux is always negative, indicating that at each location in the model conduction is a source of energy. Moreover, throughout much of the empirical atmosphere, the energy deposited by conduction exceeds the local radiative cooling rate. This forces the heating term to be positive throughout much of the transition region and corona. In other words, the heating term in the energy equation has become an energy sink.

Only below about 40,000 K ($\log T = 4.6$), is the heating term consistently a source of energy in the network model. This is because at low temperatures conduction diminishes in importance due to the $T^{5/2}$ dependence of the conductive flux. In this same temperature range, radiation becomes increasingly important because of the n_e^2 dependence in the radiative loss rate. The imbalance between radiative cooling and conductive heating is particularly severe near 10^5 K. Integrated from the $\log T = 4.7$–5.3 heights in the model, the energy loss in the heating term is about 9.9×10^5 erg cm^{-2} s^{-1}.

In the cell-center model, the divergence of the conductive flux serves as both a source and a sink for energy. At coronal temperatures it is an energy source and overwhelms the radiative losses,

Empirical Energy Balance

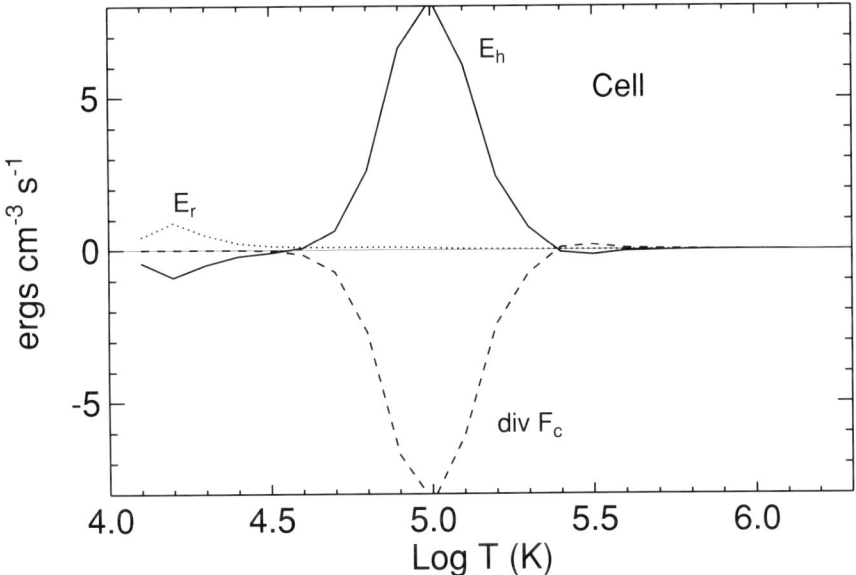

Fig. 6.6. Local cell energy balance calculated using the empirical model in Table 6.2. Positive values represent energy losses.

resulting in the heating term again becoming a sink for energy. From around 3×10^5–10^6 K, however, the conduction term is a sink for energy. At these temperatures and densities, the radiative loss rates are much lower than the divergence of the conductive flux. Thus the heating becomes a sizable source term in the empirical energy balance. Integrated between 2.5×10^5 and 10^6 K, the heating required is about 2×10^6 erg cm^{-2} s^{-1}.

Below 2.5×10^5 K, the empirical cell energy balance behaves much like the network energy balance. The divergence of the conductive flux becomes an energy source term, resulting in a region between 40,000 and 2×10^5 K in which the heating term must be a sink to achieve energy balance. The integrated energy required is about 2.8×10^6 erg cm^{-2} s^{-1}. Note that while the size of the positive peak in the heating rate at 10^5 K is much larger than the size of the negative peak in the heating rate in the upper transition region, the integrated energy totals are comparable. This is because the integration must be carried out over height and the temperature gradient is much shallower in the upper transition region. Thus, while there is large positive peak in the heating rate near 10^5 K, it does not cover a large height range in the atmosphere. Below 40,000 K, radiative losses dominate and the heating term is again an energy source.

Both the detailed local energy balance and the global energy balance depend on the pressure selected for the empirical model. If we take a $\log P_0$ of 14.8 rather than 15.0, then both the radiative loss rates and the conductive fluxes are reduced. For the radiative loss rates and the conductive flux, the reduction is proportional to n_e^2, while for the divergence of the conductive flux it is proportional to n_e^4. Thus the decrease in the divergence of the conductive flux is larger, reducing the size of the regions in the empirical models that require an additional energy loss mechanism. For example, in the network model the integrated energy emission in the large peak near 10^5 K diminishes from 9.9×10^5 to 3.5×10^5 erg cm^{-2} s^{-1}. For both the network and cell models, the downward conductive flux at 10^6 K decreases to about 5×10^5 erg cm^{-2} s^{-1}. Changing the pressure does not, however, change the integrated radiative losses from the model. This is because the local radiative loss rate decreases by a factor of n_e^2, but the thickness of the radiating layer increases, because of the decrease in the conductive flux, which decreases the temperature gradient.

This problem of the energy balance resulting from the simple static empirical models showing a need for an additional energy loss mechanism is well known. Gabriel (1976) developed an average quiet solar model following essentially the same techniques outlined here and found that the region near 10^5 K required an energy loss in addition to radiation of about 3×10^6 erg cm^{-2} s^{-1}. He suggested that the source of this problem was the assumption that the area of the emitting plasma as a function of height was constant, leading to the more sophisticated energy balance models with variable areas, which we will discuss in the next chapter.

6.5 Examining the Energy Balance Directly Using the Emission Measure

We based our analysis of the energy balance on first deriving an empirical atmospheric model and then using the model to analyze the terms in the energy balance equation to obtain the local heating rate. Jordan (1975, 1976, 1980) has developed an alternative approach in which she uses the observed emission measure gradient to derive not only the temperature and density structure of the atmosphere, but also the form of the heating function. This approach is controversial, but it does offer some insight into the nature of the energy balance in the lower transition region.

Above a temperature of about 2×10^5 K the slope of the emission measure is positive. Jordan (1980) assumed that the emission measure there could be represented by

$$\int_R n_e^2 \, dh = aT^b, \qquad (6.20)$$

where a and b are constants and R is the temperature interval over which the integral is carried out for each line. The analysis of the emission measure distribution then proceeds in the same manner as we outlined in Section 6.1, except that a linear temperature interval is used rather than a logarithmic one. This results in an expression for the temperature gradient of the form

$$\frac{dh}{dT} = \frac{\Delta T \, aT^b}{P_0^2}, \qquad (6.21)$$

where ΔT is the width of the temperature interval R and $P_0 = n_e T$. Combining this with the hydrostatic equilibrium equation then yields

$$P^2 = P_{\text{ref}} - \frac{aD}{b+1}\left(T^{b+1} - T_{\text{ref}}^{b+1}\right), \qquad (6.22)$$

where

$$D = 2 \times 0.231 \times (3.65 \mu m_p g_\odot k). \qquad (6.23)$$

The term in parenthesis in the expression for D results from the conversion from P_0 to an actual total gas pressure and we have substituted $0.231T$ for ΔT. This value is appropriate for the emission measure used in Tables 6.1 and 6.2. For a logarithmic width of 0.3 dex for the temperature, the constant would be 0.70. The reference pressure P_{ref} is evaluated at $T_{\text{ref}} = 2 \times 10^5$ K. Taking this equation to the limit where $P_0 = 0$ then defines a coronal temperature T_c given by

$$\left(T_c^{b+1} - T_{\text{ref}}^{b+1}\right) = 8.26 \times 10^{33} \frac{(b+1)}{a} P^2. \qquad (6.24)$$

Jordan (1980) used these results to analyze the simple static energy balance equation

$$\Delta F_m = \Delta F_r - \Delta F_c, \qquad (6.25)$$

where ΔF_c is the mechanical energy deposited per second in the temperature interval R, ΔF_r is the energy lost by radiation in the same temperature interval, and ΔF_c is the energy gained or lost by thermal conduction in the same interval. Using the three equations derived above, analytic fits to the radiative loss function, and the classical expression for the conductivity, Jordan was able to write an expression for ΔF_m in terms of the parameters a, b, and either P_0 or T_c.

For $b = 3/2$, she concluded that the heating mechanism must have a temperature dependence between $T^{1/2}$ and $T^{5/2}$. For all values of $b \geq 3/2$, both conduction and radiation are energy loss terms. Thus any slope in this range will require a positive heating term. For $0 < b < 3/2$, conduction can become an energy deposition term. This leads to the possibility that the energy deposited by conduction can exceed the energy radiated, resulting in a negative heating term.

Raymond and Doyle (1981b) applied this analysis to the quiet-Sun network and cell-center emission measures shown in Figure 4.1. For the cell-center emission measure they found that the observed slope of 2.1 required energy inputs of 5×10^4–8×10^4 erg cm^{-2} s^{-1} in the temperature range from 2×10^5 to 10^6 K. For the network emission measure, the observed slope of 1.4 implied an energy deposition of 3.2×10^4 erg cm^{-2} s^{-1} at 10^6 K, but energy deposition of about -10^4 erg cm^{-2} s^{-1} in the range from 2×10^5 to 5×10^5 K. This is essentially the same result that we obtained by analyzing the empirical models. Their numbers are smaller because they assumed a lower pressure.

Jordan (1980) suggested that the excess energy implied by a negative mechanical energy deposition might be the driver for spicular motions or the source of the observed nonthermal line broadening. Raymond and Doyle (1981b) interpreted the excess energy by adding a term to the energy equation describing the enthalpy flux due to flows. They found that upflow velocities of 1 km s^{-1} at 2×10^5 K could remove the excess energy. Considering the uncertainties in these numbers, however, the results are consistent with no mechanical heating or enthalpy flux in this temperature range.

This kind of analysis is subject to a several uncertainties. Just as with the normal emission measure analysis, the unknown geometry of the transition region is dealt with by assuming a simple homogeneous stratified atmosphere. In addition, representing the emission measure with an aT^b form neglects the gradual changes in the slope that are often present in individual data sets. The largest problem, however, is that, based on current observations, there are simply too many terms in the energy equation to disentangle fully using just the emission measure and the equation of hydrostatic equilibrium. Not only must radiation and conduction be considered, but also the enthalpy flux due to the downward flows seen in transition region lines. Note that Raymond and Doyle (1981b) were able to account for a derived negative mechanical energy deposition term by invoking an outflow at a temperature of 2×10^5 K. A downflow like that observed at 10^5 K would make the problem worse.

If we want to understand the energetics of the transition region, the best approach is probably to use the emission measure and either the equation of hydrostatic equilibrium or the constant pressure approximation to obtain the run of temperature, temperature gradient, and density with height, possibly with some assumptions about the geometry built in. This model can then be used to analyze the terms in the static or steady flow energy balance energy equation. When Gabriel (1976) did this, he arrived at essentially the same result as Raymond and Doyle (1981b) did for the network; the model deduced from the emission measure requires negative values of the mechanical energy deposition near 2×10^5 K.

Both Jordan (1980) and Woods et al. (1990) have applied this approach to the temperature region below the emission measure minimum. These analyses show that at temperatures below 10^5 K, the energy deposited by thermal conduction is no longer adequate to maintain the very large radiative energy losses, the same result we inferred from the energy balance analysis of the empirical atmospheric models developed in Tables 6.1 and 6.2. Thus in a static model, the additional energy must be made up through mechanical energy deposition. In fact over much of this temperature range, the static energy balance is between radiative losses and the unknown heating term.

6.6 Observed Structure and Energy Balance

It is difficult to reconcile the empirically determined atmospheric structure with the basic observations we outlined in Chapters 3, 4, and 5. While the models reproduce the observed emission measure at Sun center, they fail to explain the structuring at the limb. Observations of optically-thin emission lines at the limb show clear differences in the location of peak emission in transition region lines formed at different temperatures. As we discussed earlier, both Doschek *et al.* (1976*b*) and Mariska *et al.* (1978) found that the limb emission in lines formed above 10^5 K peaked near 4 arc sec above the white-light limb, while emission in lines formed in the 35,000–63,000 K range peaked at 2 arc sec above the white-light limb. Thus the main line forming regions in these two temperature ranges are separated by about 1450 km.

On the other hand, the height separation between 35,000 and 2×10^5 K in the network model is about 26 km. Even if we drop the scaled pressure in the models to 4×10^{14} cm^{-3} K, the spatial separation only increases to 167 km, or 0.2 arc sec. Besides showing a separation of emission peaks at the limb, observations also show that the peak is not sharp, but rather extends over 3 or 4 arc sec, compared with the tens of kilometers for the line forming region in each individual emission line predicted by the models. Combining these limb observations with disk-center intensity measurements leads to filling factors of about 1%. Obviously a significant amount of material is not distributed in height in the manner suggested by the empirical models. Faced with these clear discrepancies between the empirical models and the observed emission, it is dangerous to read a great deal into local energy balance analyses, either based on empirical models or on analyzing the slope of the emission measure curve.

What then can we learn from the empirical models? Even without considering the obvious problems with fine structure at the limb, the models show that we are missing something in the simple assumptions made in the analysis. Given the high spatial and spectral resolution observations we have already examined, the chief areas where our assumptions are likely to be flawed are the geometry and the assumption that the atmosphere is static.

While a plasma flow of a few km s^{-1} does not invalidate the hydrostatic equilibrium assumption made in constructing the empirical models (5 km s^{-1} at 10^5 K is still only 10% of the local sound speed) it does represent a major term in the energy equation through the enthalpy flux

$$F_\mathrm{f} = \frac{5}{2} Pv. \tag{6.26}$$

For a pressure of 0.2 dyn cm^{-2} and a velocity of 10 km s^{-1}, the enthalpy flux is 5×10^5 erg cm^{-2} s^{-1}. Unfortunately our knowledge of the flow field is too meager to take properly the divergence of this term in the energy equation for the empirical model.

Like the flow field, the detailed geometry in the transition region is difficult to evaluate. All the available observational evidence points to the magnetic field as the cause of the fine-scale structuring that is present in the transition region. Extrapolations of photospheric magnetic field observations to the higher layers show that much of the inner corona consists of closed loop-like structures. This is born out by soft X-ray images. Thus at transition region temperatures it is likely that there is a mixture of open and closed magnetic structures. The closed structures could connect to the surface on many spatial scales from fractions of a solar radius to the size of a supergranule cell or an individual patch of UV emitting material. Thus there are many possible geometries to examine. Observations do not yet provide sufficient information to determine the proper geometry. While the diverging geometry suggested by Gabriel (1976) has dominated our ideas about the structure of the transition region, it is not the only possibility for the plasma distribution.

There is the possibility that our understanding of the geometry of the solar transition region is completely wrong. Feldman (1983, 1987) for example has argued that most of the plasma in the temperature range from 20,000 to 5×10^5 K resides in unresolved fine structures that are isolated from the chromosphere and corona. He based this conclusion on the observed behavior of transition region lines above the limb as shown for example in Figure 3.5; the turnover in nonthermal velocity measurements at coronal temperatures; and observations suggesting chromospheric and coronal densities are roughly constant in coronal holes, quiet Sun, and active regions, while the transition region density varies by more than an order of magnitude. Only the first argument is truly compelling, but they all remind us that we should be willing to consider radical solutions to explain the observations. In the next chapter we consider both conventional models and some newer models.

7
Physical Transition Region Models

The physics of the transition region is complex. Simple empirical models, such as those we developed in the last chapter, suggest that thermal conduction is a major factor in determining the temperature and density structure. Measurements of UV and EUV line fluxes show that the amount of energy radiated from the transition region is about 1.9×10^5 erg cm^{-2} s^{-1} (Kopp, 1972), an appreciable fraction of the estimated conductive flux. In addition, the flow measurements outlined in Chapter 5 yield local enthalpy fluxes of 4×10^5 erg cm^{-2} s^{-1}. The measured nonthermal broadening discussed in Chapter 5 also represents a potential source of energy. Finally, both the fine structure observed in the transition region and the large-scale organization of the corona, show clear evidence for structuring imposed by the magnetic field. To make further progress in understanding this region of the solar atmosphere, we must examine the physics implied by the observations in more detail. The goal of this discussion is to develop a theoretical model that describes the structure of the transition region and relates it to the flow of mass, momentum, and energy through the outer layers of the solar atmosphere.

7.1   Role of the Magnetic Field

While our goal is to develop a model for the thermodynamic properties of the transition region, it is clear from the observed structuring that the magnetic field plays a role. Analyses of visible and infrared wavelength observations show that magnetic flux emerges from beneath the visible surface in 1–2 kG discrete elements with sizes well under 1 arc sec (e.g., Stenflo, 1989). Regions between these elements have little or no magnetic field.

One way to characterize the relationship between the plasma and the magnetic field is to calculate the plasma β, where

$$\beta = \frac{4\pi P}{B^2}, \tag{7.1}$$

is the ratio of the gas pressure to the magnetic pressure. The magnetic field shapes the plasma structure when $\beta \lesssim 1$. At photospheric levels the magnetic pressure in the magnetic elements is much larger than the gas pressure and $\beta < 1$. Between the magnetic elements the gas pressure dominates and $\beta \gg 1$. As we move outward in the solar atmosphere, the gas pressure drops exponentially, while the magnetic pressure drops more slowly. Thus the β of the regions outside the discrete magnetic elements decreases rapidly and the magnetic elements expand. In the corona $\beta \ll 1$ and the magnetic field fills the entire volume.

The magnetic field in the transition region and corona is governed by Ampère's law,

$$\nabla \times \mathbf{B} = \frac{4\pi}{c} \mathbf{j}, \tag{7.2}$$

Faraday's law,

$$\nabla \times \mathbf{E} = -\frac{1}{c} \frac{\partial \mathbf{B}}{\partial t}, \tag{7.3}$$

Gauss' law,

$$\nabla \cdot \mathbf{B} = 0, \tag{7.4}$$

and Ohm's law

$$\eta \mathbf{j} = \mathbf{E} + \frac{\mathbf{v} \times \mathbf{B}}{c}, \tag{7.5}$$

where all the equations are in Gaussian units. Since we are primarily concerned with the magnetic field, it is convenient to eliminate the electric field $\mathbf{E}$ and the current density $\mathbf{j}$ from equations (7.2), (7.3), and (7.5) to give

$$\frac{\partial \mathbf{B}}{\partial t} = \nabla \times (\mathbf{v} \times \mathbf{B}) - \frac{\eta c^2}{4\pi} \nabla \times (\nabla \times \mathbf{B}), \tag{7.6}$$

where we have assumed that the plasma resistivity η is constant. Using the vector identity

$$\nabla \times (\nabla \times \mathbf{B}) = \nabla(\nabla \cdot \mathbf{B}) - \nabla^2 \mathbf{B}, \tag{7.7}$$

and Gauss' law, equation (7.6) becomes

$$\frac{\partial \mathbf{B}}{\partial t} = \nabla(\mathbf{v} \times \mathbf{B}) + \frac{\eta c^2}{2\pi}\nabla^2 \mathbf{B}, \tag{7.8}$$

which is known as the induction equation or the magnetic diffusion equation. For a given velocity $\mathbf{v}$, this equation, along with Gauss' law, describes how the magnetic field evolves in time. Once we know the magnetic field, the current density and electric field follow from Ampère's law and Ohm's law. The first term on the right side of the equation describes how plasma motions convect the magnetic field, while the second term describes how resistive diffusion destroys magnetic field.

For a given $\mathbf{v}$, the relative importance of the two terms in the induction equation determines how the magnetic field will evolve. The ratio of the magnitude of the convective term in equation (7.8) to the diffusive term is defined as the magnetic Reynolds number and is approximated by

$$R_m = \frac{4\pi v l}{\eta c^2}, \tag{7.9}$$

where l is the scale length for variation of B. For the solar plasma in the transition region or corona (e.g., Spitzer, 1962), $\eta \approx 8 \times 10^8 \ln \Lambda\, T^{-3/2}$ esu, where $\ln \Lambda$ is the Coulomb logarithm. Thus

$$R_m \approx 1.8 \times 10^{-12} \frac{v l T^{3/2}}{\ln \Lambda}. \tag{7.10}$$

In the transition region, the relevant velocity is the Alfvén speed

$$V_A = \frac{B}{(4\pi\rho)^{1/2}} = 2.2 \times 10^{11} B n_e^{-1/2}. \tag{7.11}$$

Taking $T = 10^5$ K, $n_e = 10^{10}$ cm^{-3}, and $\ln \Lambda = 10$, the magnetic Reynolds number becomes

$$R_m \approx 12.5 B l. \tag{7.12}$$

For a typical 3 arc sec (about 2×10^8 cm) observed structure size in the transition region, $R_m \gg 1$ for any reasonable value of B. Even for a filling factor of 0.01, $R_m \gg 1$. In the corona where the characteristic length scales are even larger, the same result holds.

Thus for the observed structures in the transition region and corona, only the convective term in the induction equation is important. Taking the large magnetic Reynolds number result to the extreme case of a perfectly conducting plasma leads to the important result that the magnetic flux through any surface bounded by a closed contour moving with a perfectly conducting fluid is constant.

In addition, a plasma element that initially lies on a magnetic field line will continue to lie along that field line. Proofs of these results can be found in Boyd and Sanderson (1969). These results can be understood physically by considering Ohm's law and Faraday's law (e.g., Tandberg-Hanssen and Emslie, 1988). For a perfectly conducting fluid, Ohm's law implies that a finite **E** would drive an infinite current, which is impossible. Faraday's law says that **E** is produced by a circuit element cutting through a line of force. Since **E** must be zero, the plasma must move with the lines of force.

For an element of gas in a gravitationally stratified atmosphere, the inward force of gravity balances the outward directed pressure force. In the presence of a magnetic field, there is in addition the Lorentz force. The hydrostatic equilibrium equation then becomes

$$\nabla P = \rho \mathbf{g} + \mathbf{j} \times \mathbf{B}. \tag{7.13}$$

In the transition region, however, we have already shown that the force of gravity may be neglected since the structures are typically smaller than the pressure scale height. Moreover, the fact that the plasma β is much smaller than unity implies that any pressure gradient in equation (7.13) is much smaller than the Lorentz force. Equation (7.13) then becomes

$$\mathbf{j} \times \mathbf{B} = 0, \tag{7.14}$$

and magnetic fields satisfying it are referred to as force free. Physically, this equation says that for a static configuration a finite Lorentz force cannot be present because there are no other forces present to balance it. This equation can be satisfied if $\mathbf{B} = 0$, if $\mathbf{j} = 0$, or if $\mathbf{j}$ is parallel to $\mathbf{B}$. In the second case the resulting magnetic field is referred to as current free or potential.

Of course the magnetic field is not static. It evolves on time scales of hours to days, which are long compared with many dynamic phenomena we observe in the transition region. Taken together then, the features of the solar magnetic field presented here imply that the structure of the magnetic field is independent of the plasma distribution in the transition region and that plasma motions are constrained to follow the magnetic field structure. Thus the magnetic field provides the geometry in which the equations describing mass, momentum, and energy conservation for the plasma must be solved.

While it may at first appear that the magnetic field serves only to channel the plasma motions, it also may play a major role in maintaining the transition region. As we saw in Chapter 5, despite careful searches, there is no evidence for acoustic wave heating in the transition region and corona. This leaves Alfvén waves and the dissipation of electric currents as prime candidates for the transition region heating source. Alfvén waves could easily be generated by mass motions lower in the atmosphere, but require a dissipation mechanism to deposit their energy. Similarly mass motions lower in the atmosphere can produce nonpotential magnetic fields in the corona. The problem again is how to extract that energy.

7.2 Model Equations

In the transition region the density is high enough that electrons and ions collide frequently. Thus for all reasonable time scales for quiet-Sun phenomena we can treat the plasma as a single fluid with a total pressure P, mass density ρ, and temperature T. The channeling effect of the magnetic field means that the conservation equations for these variables need to be solved in only one dimension. Thus if we take s to be the position measured along a magnetic field line and $A(s)$ to be the cross-sectional area of the magnetic flux tube that the field line is part of, the equations for mass, momentum, and energy conservation become

$$\frac{\partial}{\partial t}(\rho A) + \frac{\partial}{\partial s}(\rho v A) = 0, \tag{7.15}$$

$$\frac{\partial}{\partial t}(\rho v A) + \frac{\partial}{\partial s}(\rho v^2 A) = -A\left(\frac{\partial P}{\partial s} + \rho g_\|\right), \tag{7.16}$$

and

$$\frac{\partial}{\partial t}(EA) + \frac{\partial}{\partial s}(EVA) = -\frac{\partial}{\partial s}(PvA) + \frac{\partial}{\partial s}\left(A\kappa\frac{\partial T}{\partial s}\right) + \rho v A g_\| + A(H - L), \tag{7.17}$$

where

$$E = \frac{1}{2}\rho v^2 + \frac{P}{\gamma - 1}. \tag{7.18}$$

In these equations $g_\|$ is the gravitational acceleration along the magnetic field; γ is the ratio of specific heats, usually taken to be 5/3; κ is the thermal conductivity; L is the energy loss rate due to radiation; and H is the heating rate, which may be a complicated function of the other variables.

Completing this set of equations is the perfect gas law

$$P = \frac{\rho}{\mu m_\text{p}} kT. \qquad (7.19)$$

Above a temperature of about 10^5 K, we can treat the plasma as fully ionized. Below about 20,000 K atomic hydrogen and helium are present in significant quantities (e.g., McWhirter et al., 1975). For a fully ionized plasma consisting of hydrogen and 10% helium by number, $\mu = 0.61$. The total particle number density is then

$$n = \frac{\rho}{\mu m_\text{p}} = n_\text{c} + n_\text{p} + n_\text{He} \qquad (7.20)$$

and the electron number density is

$$n_\text{e} = n_\text{p} + 2 n_\text{He}. \qquad (7.21)$$

Substituting for the helium abundance, we have $n_\text{p}/n_\text{e} = 0.83$ and $n = 1.91 n_\text{e}$. For rough calculations we can usually assume that $n = 2 n_\text{e}$.

Many processes contribute to the heat flux along the magnetic field $\kappa \, \partial T / \partial s$. Thermal energy can be transported by electrons, ions, and neutrals, all of which may collide with themselves or each other. Ulmschneider (1970) has calculated the coefficient of thermal conductivity for pure hydrogen, pure helium, and a plasma with cosmic abundances. Above a temperature of about 20,000 K where the plasma is fully ionized, the conductivity agrees with the expression for the electron conductivity used by Spitzer (1962). Thus we can take

$$\kappa = 10^{-6} T^{5/2} \qquad \text{erg s}^{-1} \text{ K}^{-1} \text{ cm}^{-1}. \qquad (7.22)$$

The ion conductivity coefficient is only about 4% of the electron conductivity coefficient (e.g., Braginskii, 1965) and thus may be ignored. While the relative contribution of neutrals to the total thermal conductivity becomes important at low temperatures (e.g., Ulmschneider, 1970), the absolute value is small enough that the conduction term in the energy equation ceases to be important.

Along the magnetic field conduction is primarily due to electrons. Across the magnetic field that is not true. Since the charged particles are constrained to spiral along the field lines, the conductivity perpendicular to the magnetic field depends on the inverse of the product of the square of the gyrofrequency and the collision time.

Because of their large mass and large radius of gyration, protons dominate the transverse thermal conductivity. The relative importance of the parallel and perpendicular conductivities is given by the ratio of the perpendicular ion conductivity coefficient to the parallel electron conductivity coefficient. From Spitzer (1962) we have

$$\frac{\kappa_\perp}{\kappa_\parallel} = 8 \times 10^{-14} \frac{n_e^2 (\ln \Lambda)^2}{T^3 B^2}. \quad (7.23)$$

Taking typical transition region values of 10^{10} cm^{-3} for the electron density, 10^5 K for the temperature, and 10 for $\ln \Lambda$, gives a value for the ratio of $8 \times 10^{-7} B^{-2}$. Thus we can neglect thermal conductivity across the magnetic field.

As we showed in our examination of the empirical atmospheric models, radiation is a major energy loss mechanism in the transition region. Throughout most of the transition region, the radiation loss rate is given by

$$L = n_e n_p P(T), \quad (7.24)$$

the optically-thin expression we discussed in Chapter 2. In the 10,000–30,000 K range, radiation from the hydrogen Lyman α line dominates (e.g., Vernazza et al., 1981). The radiative loss rates then depend on the details of the atmospheric structure, particularly the hydrogen ionization. Thus at low temperatures the nature of the modelling problem changes dramatically. Instead of a set of fluid dynamics equations describing mass, momentum, and energy conservation, we are confronted with the additional complication of the full non-LTE radiative transfer problem. That problem is beyond the scope of this book.

Even with the assumption that the radiative losses can be described by an optically thin steady state treatment, the full set of mass, momentum, and energy conservation equations represents a formidable challenge. In principle once we know the geometry, the boundary and initial conditions, and the nature of the heating term, both its spatial and temporal variation, we can solve the equations to yield a time-dependent description of the thermodynamic variables. We can then use this description to predict the emission in transition region lines for comparison with observations. In practice we do not have a full understanding of the required initial and boundary conditions and some terms in the energy equation. In particular, we know little about the heating term and are reduced to studying the implications for the atmospheric structure of various assumptions about it.

7.3 Static Models

If the plasma in the transition region is static, the mass conservation equation is satisfied trivially, the momentum conservation equation reverts to the hydrostatic equilibrium equation with $g_\parallel$ possibly varying with position because of the magnetic field, and the energy equation becomes

$$-\frac{1}{A(s)}\frac{\mathrm{d}}{\mathrm{d}s}\left[A(s)\kappa_0 T^{5/2}\frac{\mathrm{d}T}{\mathrm{d}s}\right] = H - L. \tag{7.25}$$

With suitable boundary conditions and an expression for the heating rate, we can easily integrate the hydrostatic equilibrium equation and the energy equation to yield a temperature and density model for the transition region and corona. With this model we can then calculate a theoretical emission measure curve to compare with observational data.

7.3.1 Scaling Laws

In a loop geometry, we can use equation (7.25) to obtain estimates of how the atmospheric parameters will behave. For loops less than about 5×10^9 cm in height, a coronal scale height, we can assume that the pressure is a constant and need only to work with the energy equation. Consider a magnetic loop with a constant cross-sectional area, length from base to apex L, base temperature T_0, and apex temperature T_m. Taking the heating rate as constant per unit volume, we can write equation (7.25) as

$$\frac{\mathrm{d}F_c}{\mathrm{d}s} = n_e^2 \chi T^{-1/2} - H, \tag{7.26}$$

where we have taken a $T^{-1/2}$ temperature dependence for the radiative loss function over the entire transition region and low corona and $\chi = 10^{-14.09}$ (e.g., Priest, 1982, p. 89). Multiplying equation (7.26) by F_c and removing n_e using the perfect gas law, we have

$$F_c\,\mathrm{d}F_c = \frac{P^2 \kappa_0 \chi}{4k^2}\,\mathrm{d}T - H\kappa_0 T^{5/2}\,\mathrm{d}T, \tag{7.27}$$

which integrates to

$$\frac{1}{2}F_c^2 = \frac{P^2\kappa_0\chi}{4k^2}(T - T_m) - \frac{2H\kappa_0}{7}(T^{7/2} - T_m^{7/2}). \tag{7.28}$$

We have assumed that the temperature gradient is zero at the top of the loop, so that F_c at T_m is also zero.

At the base of the loop, the conductivity is small. Taking the limit of a thermally isolated loop, we set it to zero at T_0 and obtain

$$H = 3.5\frac{P^2\chi}{k^2}T_m^{-5/2}, \tag{7.29}$$

where we have assumed $T_m \gg T_0$. Substituting this expression for H in equation (7.28), we have

$$T^2\frac{dT}{ds} = \left(\frac{P^2\chi}{2k^2\kappa_0}\right)^{1/2}\left(1 - \frac{T^{5/2}}{T_m^{5/2}}\right)^{1/2}. \tag{7.30}$$

To within an order of magnitude, we have

$$T_m^3 \approx \left(\frac{\chi}{2k^2\kappa_0}\right)^{1/2}PL, \tag{7.31}$$

a scaling law for the relation between the peak temperature and the pressure and loop length. Carrying out the integral yields

$$T_m \approx 1.4 \times 10^4(PL)^{1/3}. \tag{7.32}$$

Substituting T_m from equation (7.31) in equation (7.29) for H we have

$$H \sim P^{7/6}L^{-5/6}. \tag{7.33}$$

Combining this equation with equation (7.31) results in expressions relating the temperature to the heating rate and the loop length and the pressure to the heating rate and the loop length.

Rosner et al. (1978) and Craig et al. (1978) first developed these scaling laws for thermally isolated loops. Data for many active region loops appear to follow equation (7.32). For quiet loops, however, we have less information about the geometric structure. Taking a characteristic temperature of 10^6 K and a pressure of 0.2 dyn cm^{-2} yields a length of about 18,200 km, about half the size of a supergranule cell. For a temperature of 2×10^6 K, the length is 1.45×10^5 km, many supergranule cells in size. If the structure in an individual patch of network consists of thermally isolated loops, then the characteristic size scale of about 7250 km (~ 10 arc sec) and a pressure of 0.2 dyn cm^{-2} yield maximum temperatures of about 7×10^5 K. Taking this to the extreme of the 3 arc sec size reported by Dere et al. (1987) gives peak temperatures of only about 4.9×10^5 K, suggesting that many loop like features in the transition region network may not reach coronal temperatures.

7.3.2 Static Model Solutions

There have been many studies of the solutions to equation (7.25). Since conduction and radiation are well understood, work has concentrated on the effects of different geometries and heating functions on the solutions. Early solutions considered plane-parallel (Moore and Fung, 1972) and spherical geometries (McWhirter et al., 1975). Gabriel (1976) first considered expanding magnetic field geometries like those that may be present in the network. More recent analyses have generally been for loop structures that have parameters typical of the coronal structures observed in X-rays to be associated with active regions. Thus their base pressures are a few dyn cm^{-2}, instead of the roughly 0.2 dyn cm^{-2} typical of the quiet solar transition region.

Many studies have concentrated on the heating term. Because empirical models for the transition region suggest that it is narrow and conduction dominated, the simplest assumption that we can make about the heating term is that all the energy is deposited well above the transition region and appears as a thermal conductive flux at the top of the model. Thus the heating term disappears from the energy equation and a finite conductive flux boundary condition at the top of the model replaces it. Calculations of this type have been performed by Moore and Fung (1972), Gabriel (1976), and McWhirter et al. (1975). Models have also been constructed with the heating constant per unit volume (Vesecky et al., 1979), with the heating deposited as an exponential over height with some characteristic scale height (Pallavicini et al., 1981; Serio et al., 1981), and with the heating proportional to the temperature to some power (Chiuderi et al., 1981; Torricelli-Ciamponi et al., 1982).

Calculating families of static models with various assumptions about the boundary conditions, geometry, and heating rate is simple. Comparing those calculations with quiet-Sun observations is much more of a challenge. As an example, consider the case of an energy deposition rate that is constant per unit volume. If we assume a simple vertical flux tube of constant cross-sectional area with specified base pressure, temperature, conductive flux,

206 *Physical Transition Region Models*

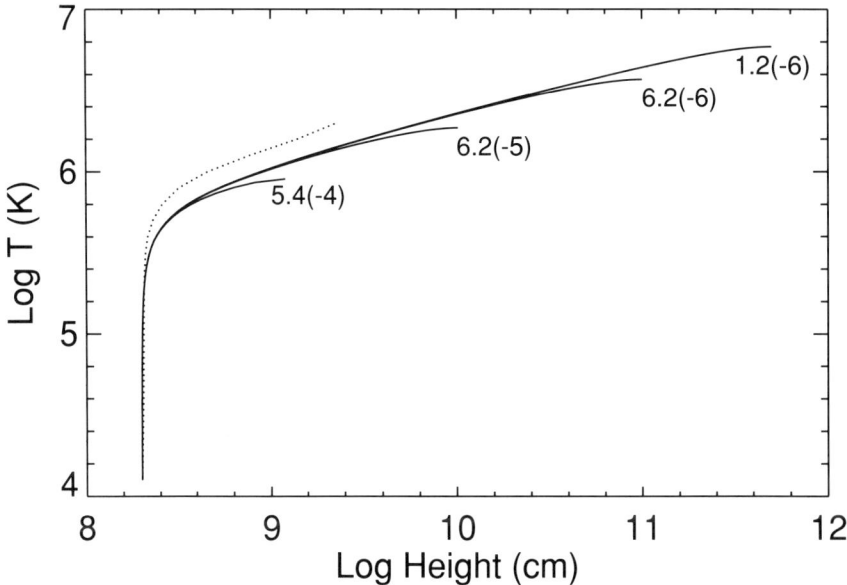

Fig. 7.1. Temperature as a function of height for atmospheric energy balance models with a base pressure of 10^{15} cm^{-3} K and the constant volumetric heating rates shown. The dotted line is the network empirical model listed in Table 6.1.

and value for the heating rate, then we can integrate the hydrostatic equilibrium and energy equations upward until we reach a maximum temperature. We can then compute the model emission measure and compare it with the observed emission measure to assess the degree to which the models agree with the observations.

Figure 7.1 illustrates the first step in this process. Here we plot the temperature as a function of height for a series of models with base pressures of $10^{15.0}$ cm^{-3} K and the base temperature and conductive flux listed in Table 6.1 for the network empirical model. The different models are characterized by the height of the temperature maximum, which depends on the heating rate. Also plotted on the figure as the dotted line is the network empirical model from Table 6.1.

As we expect based on our energy balance analysis of the empirical models, there are significant differences between the empirical model and the energy balance models. All the energy balance models do not reach coronal temperatures until much greater heights

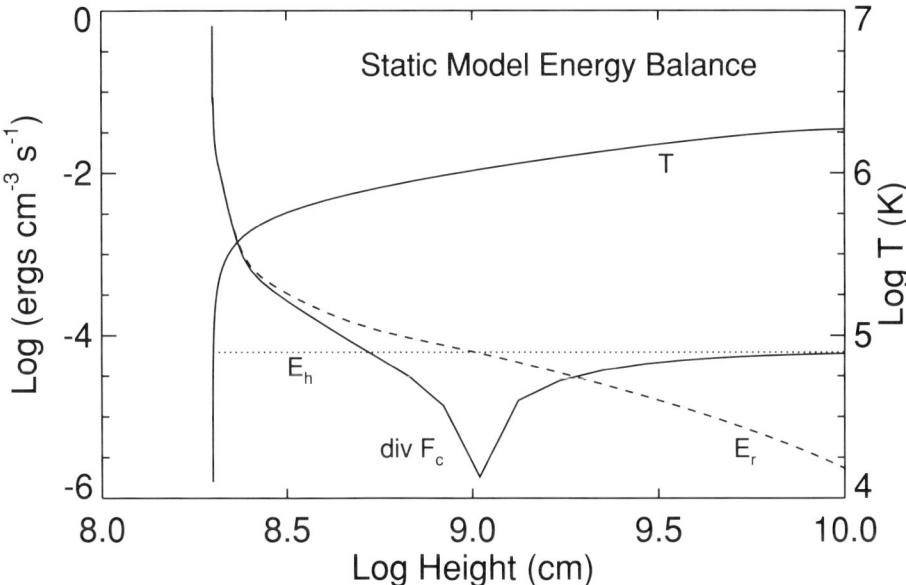

Fig. 7.2. Theoretical energy balance for the model shown in Figure 7.1 with a maximum height of 10^{10} cm. For reference the model temperature distribution is also plotted.

than the empirical model. In addition, the conductive flux in the transition region, the other parameter that we can easily determine from a simple empirical model based on the emission measure, does not agree. While the empirical model provides an estimate of the conductive flux in the network of about 10^6 erg cm^{-2} s^{-1} at 3×10^5 K, the energy balance models have a conductive flux at the same temperature of about 5×10^4 erg cm^{-2} s^{-1}. Moreover, instead of remaining constant, the conductive flux decreases as the temperature decreases.

This change in the conductive flux with temperature is due to the requirement that the theoretical model be in energy balance. Figure 7.2 illustrates how the energy balance in a simple static model works. Here we plot the absolute values of the terms in the energy equation as a function of height in the model for the static model with a maximum length of 10^{10} cm. Since the energy deposition in this model is constant per unit volume, the heating rate is a horizontal line, which has a value of 6.2×10^{-5} erg cm^{-3} s^{-1}. In the corona the radiation rate is low because of both the low density and the temperature dependence of the optically-thin radiative loss rate. Thus conduction removes the energy deposited by the heating and the divergence of the conductive flux is a loss term in the energy equation.

Throughout the inner corona the conduction term continues to remove energy deposited by the heating. As we move inward, however, the temperature drops and the density increases, leading to increasing radiative losses. Thus the magnitude of the conductive losses decreases as the importance of radiation increases. At the top of the transition region, at a height of about 10^9 cm and a temperature of 10^6 K in this model, the energy deposited by the heating mechanism exactly balances the energy lost to radiation and the divergence of the conductive flux is zero. Below this point there is more radiation than heating, and the divergence of the conductive flux changes from an energy loss term to an energy source term. Finally in the low transition region, the heating is small compared with the radiation and conduction terms, and the energy conducted from above becomes the primary source to power the radiative losses.

In these simple energy balance models, conduction plays a complicated role in the physics of the transition region and corona. It can be either a source of energy or a sink for energy. When we consider the entire model, however, it is clear that the only role that conduction plays is to redistribute energy within the system. If the temperature gradient at the top of the model is zero and the conductive flux out of the bottom of the model is small or even zero, then the only way to put energy into the system is through the heating term and the only way to remove energy from the system is by radiation.

If we assume that the area of the flux tube remains constant, it is difficult to bring the theoretical models into closer agreement with the empirical ones. As we noted earlier, for a loop geometry the static energy balance equation implies that the pressure, loop length, and peak temperature are related. Thus for a given pressure, the only way to increase the temperature is to increase the loop length. As the models in Figure 7.1 show, however, the temperature rise will continue to be too shallow. The only way to increase the temperature gradient and improve the fit with the empirical model is to increase the pressure. Density diagnostics are,

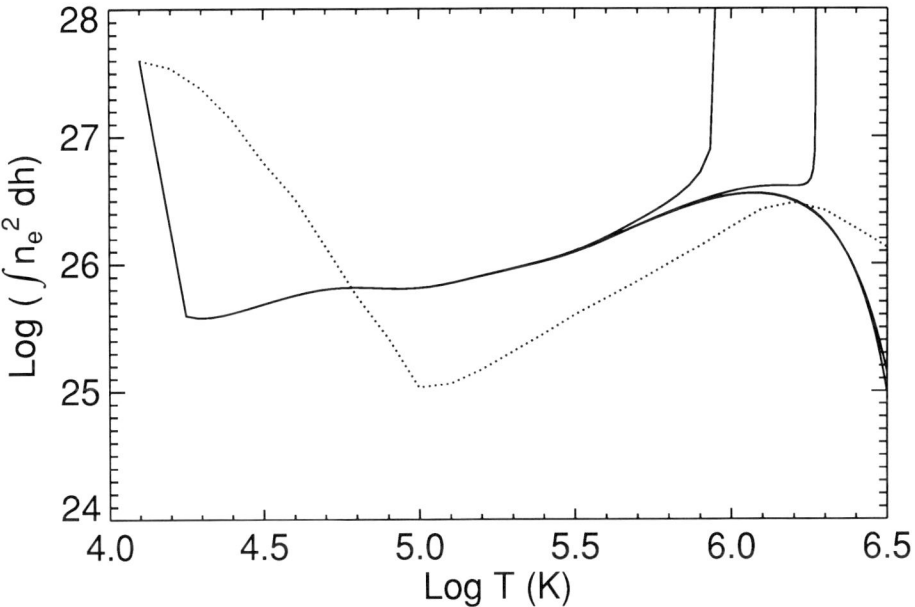

Fig. 7.3. Emission measures for the theoretical energy balance models shown in Figure 7.1. The dotted line is the empirical network emission measure listed in Table 6.1.

however, well developed and are independent of assumptions about the geometry. Thus the model pressure cannot be varied by more than about a factor of 2. This range of variation is not sufficient to improve significantly the agreement between the energy balance model and the empirical model.

7.3.3 Emission Measures

A more direct way to compare the model calculations with the observations is to calculate the emission measure at each temperature in the model and compare it with the empirical emission measure. Figure 7.3 shows this comparison. Care must be taken when placing the theoretical emission measure on the same scale as the empirical result. Typically the quantity available directly from the model is $n_e^2(\mathrm{d}T/\mathrm{d}h)^{-1}$ at each height in the model. For comparison with the emission measure curve derived by Raymond and Doyle (1981b), the $\mathrm{d}T/\mathrm{d}h$ must be converted to $\mathrm{d}\log T/\mathrm{d}h$ and the entire expression multiplied by the $\Delta \log T$ interval used in

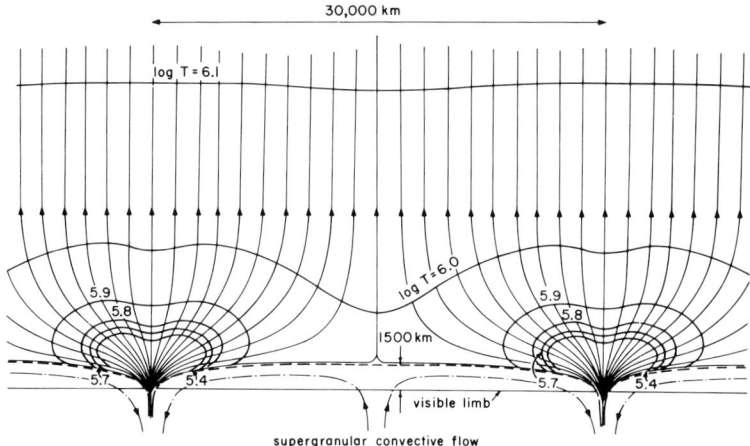

Fig. 7.4. Magnetic field geometry suggested by Gabriel (1976).

the empirical determination. Once we have made this conversion, there are no adjustments available that will change the absolute value of the computed emission measure. The sharp initial drop in the emission measure models is due to the sudden increase in the temperature gradient that is required to produce enough conductive energy input to balance the very large radiative energy losses at the base of the model.

As Figure 7.3 shows, the agreement in absolute values is excellent at the coronal peak in the emission measure. The small difference at the peak in the empirical curve is well within the range that can be accounted for by a small decrease in the pressure in the model. Below the coronal emission measure peak, however, the fit is poorer because the slope of the theoretical curve is smaller than the empirical value. Gabriel (1976) pointed out this inability of the emission measure derived from simple static energy balance models to match the observed slope. He argued that the observational evidence suggested that the magnetic field expanded with height and that this expansion must be accounted for in an energy balance model. When this was included, the slope of the derived emission measure steepened, improving the fit with the observational values.

Figure 7.4 shows the magnetic geometry obtained by Gabriel. This geometry is based on the assumption that the magnetic field at photospheric levels is concentrated by the supergranule flow into network elements and then expands at higher levels until it finally fills the corona. It is only obtained when the magnetic field at photospheric levels is unipolar over an area at least as large as a supergranule cell. As we discussed in Chapter 3, high-resolution magnetic field observations suggest that the small-scale magnetic field is in fact composed of bipolar structures that are much smaller in size than a supergranule cell. Thus it is not clear how often the geometry shown in Figure 7.4 is present on the Sun.

Dowdy et al. (1987) have examined the ability of static energy balance models in diverging magnetic geometries to produce emission measure curves that match the observations. They assumed that a downward conductive flux from the corona provided all the transition region heating. Thus their energy equation contained only conduction and radiation. Assuming a constant pressure, they specified an upper and lower boundary temperature and either a flux tube length or conductive flux at the lower boundary. The downward coronal conductive flux was thus determined by the model.

They found that for simple cone shaped geometries only models with constriction factors near 3 between the 10^6 K level and the 24,000 K level produced rough agreement with the observed emission measure for temperatures greater than about 2×10^5 K. For geometries with most of their expansion near the corona, the calculated emission measure decreased too rapidly with decreasing temperature to match the observations. For geometries in which most of the expansion occurred near the base of the transition region it was possible to match the observed emission between 2×10^5 and 10^6 K. Dowdy et al. (1987) argued, however, that the observation that the network occupies no more than about 0.45 of the solar surface up to a temperature of 7×10^5 K (Reeves, 1976) precluded those geometries. They further argued, based on the work of Dowdy et al. (1986), that the expansion of the magnetic field from the base of the network transition region to the corona must be of order 100. This plus the restriction on the fractional area covered by the network at 7×10^5 K then leads to the conclusion that none of the simple conductively heated models will satisfy all the observational constraints between 2×10^5 and 10^6 K.

Rabin (1991) examined extensively the energy balance in model transition regions confined within magnetic geometries with a wide range of shapes and lengths. Like Dowdy et al. (1987), he concluded that models with a bowl shaped geometry and small constriction factors, about a factor of 4, provided the best fit to the observations above a temperature of 2×10^5 K. His model also satisfied the network area constraint. No models with larger magnetic field expansion factors satisfied the observations.

7.4 Steady Flow Models

While Gabriel's model provided an improved fit to the emission measure above 10^5 K, it, and all the models similar to it, failed to predict the emission measure at lower temperatures. Moreover, high spectral resolution observations began to show that downflows are ubiquitous in the transition region. Pneuman and Kopp (1977) showed that these downflows represented a significant term in the energy equation, and thus needed to be considered in developing numerical models.

The energy equation for a low-velocity steady flow must include most of the terms in equation (7.17). Although the transport of kinetic energy by the flow is small, the transport of internal energy is sizable. In addition the first term on the right side of equation (7.17), which represents the work done by the flow on the surrounding plasma, must be included. Taken together these two terms constitute the enthalpy flux. In addition, we also must include the gravitational potential energy. For a steady flow the energy equation is then

$$-\frac{1}{A(s)}\frac{d}{ds}\left[A(s)\kappa\frac{dT}{ds}\right] = H - L + \rho v A(s) g_{\parallel} \\ -\frac{1}{A(s)}\frac{d}{ds}\left[\frac{\gamma P v A(s)}{\gamma - 1}\right]. \quad (7.34)$$

For a steady downflow both the enthalpy and the gravitational potential represent sources of energy.

7.4.1 Steady Downflow Models

Many authors have examined solutions to this steady flow equation. Pneuman and Kopp (1978) ignored the conduction term and investigated solutions in which the enthalpy flux supplied the energy for the transition region radiative losses. With no change in the cross-sectional area with height, this model provided satisfactory agreement with the observed emission measure distribution for temperatures above about 2×10^5 K, but not at lower temperatures. Athay (1981, 1982) analyzed more realistic energy balance models that included conduction, mass flow, and gravitational potential energy as energy sources to balance radiation in the transition region. He also included a detailed consideration of possible magnetic field geometries. Athay found that it was possible to match the observed emission measure distribution for temperatures greater than about 2×10^5 K with a variety of models. None of the models he studied was able to match the emission measure distribution at lower temperatures.

Both Pneuman and Kopp (1978) and Wallenhorst (1982) were able to extend this agreement to lower temperatures by making the downward particle flux decrease with increasing temperature. This result is based on the idea that an upward spicule flux is the source of the downflow and is in rough agreement with the observation that the number of spicules declines with increasing height above the limb. It does not, however, address the coupling between the upflowing plasma and the downflowing plasma.

Fiedler and Cally (1990) extended earlier work on flows in expanding magnetic geometries by including the magnetic terms in the momentum and energy equations. The resulting geometry is essentially the same as that used by Gabriel (1976) and Athay (1982). As was true in the earlier studies, Fiedler and Cally concluded that they could find solutions to their equations that matched the observed emission measure above 10^5 K. The models, however, generally had velocities at 10^5 K that were less than 1 km s^{-1}, much smaller than the observed downflows. For models with classical thermal conductivity, there were no solutions that matched the observed emission measure distribution below 10^5 K.

7.4.2 Loop Flow Models

All the steady flow models discussed above considered only the transition region and downflowing plasma. The modeler assumed that spicular material is ejected into the corona where it is heated to coronal temperatures and returns as the observed downflowing plasma. Some efforts have been made to understand a more complete system that accounts for both the upflow and downflow.

While there are no studies of the dynamics of spicule upflow followed by a steady downflow of material, there have been several investigations of steady flows in loop structures. Early studies in this area (Cargill and Priest, 1980; Noci, 1981) assumed a polytropic flow, and therefore the results are of questionable validity for understanding flows on the Sun (Yeh, 1977).

Yeh (1977), Cargill and Priest (1982), and Antiochos (1984) investigated steady state models with a more realistic energy equation. While these studies do provide some insight into the physics of flows in magnetic loops, they failed to include properly the dynamic coupling that is present between the chromosphere, transition region, and corona. Instead they produced the flow by introducing a pressure difference between the two boundaries in the calculation. This is mathematically acceptable, but fails to address the physical question of how to maintain such a pressure difference. In the Sun there is no real boundary at the base of the transition region. Any disturbance in the pressure will quickly return to an equilibrium with only transient flows produced. The fundamental problem with the observed steady flows on the Sun is, what drives them? The best way to answer that question is with time-dependent calculations that place the computational boundary conditions a significant distance away from the transition region.

Boris and Mariska (1982) and Mariska and Boris (1983) considered a more realistic loop model that included an idealized chromospheric region at each end. They solved the full set of time-dependent mass, momentum, and energy equations for a semicircular loop with a constant cross-sectional area. In their model an asymmetry in the energy deposition drove the flow. This eliminated the need for artificially imposed boundary conditions. Beginning with a static initial atmosphere, they gradually changed the spatial dependence of the heating term in the energy equation from the uniform volumetric energy deposition to one in which 50% of the total energy was deposited in a small region on one side of the loop. This heating asymmetry drove a quasisteady flow from one end of the loop to the other with a peak velocity of roughly 5 km s^{-1}. The flow velocity at the temperature of formation of the C IV resonance lines was, however, smaller than the observed values of 5–10 km s^{-1}.

Craig and McClymont (1986) extended these results with a detailed examination of transition region models with mass flow. They found that the emission measure of the downflowing plasma was greater than the upflowing plasma, in agreement with the observed predominance of downflows. For modest changes in the coronal heating rate, however, they were unable to produce flows that were fast enough in the temperature range of formation of the C IV resonance lines to match the observed downflow velocities. Moreover, like the simulations of Boris and Mariska (1982) and Mariska and Boris (1983), they found that the peak redshift should occur at higher temperatures than the temperature of formation of C IV. The available observational data suggest that the flows are no longer present at higher temperatures.

Further analyses of loop flows driven by heating asymmetries by McClymont and Craig (1986, 1987) showed that it is possible to produce flow velocities of the required magnitude. They found that as the peak loop temperature increased, the flow velocity at 10^5 K would decrease. This is a natural consequence of the conservation of momentum and implies that the loops responsible for the observed flows must be relatively cool. For velocities at 10^5 K of 5–10 km s^{-1}, McClymont and Craig (1986) found that the peak temperature in the loop must be in the range 4×10^5–1.6×10^6 K. If the flows are in quiet loops the pressure must be about 0.2 dyn cm^{-2}. Using the Rosner et al. (1978) scaling law immediately implies that the total loop length should be in the range from 1200 to 75,000 km, with the small loops having the largest velocities.

McClymont and Craig (1986, 1987) also found that asymmetric heating could drive flows of the required speed if nearly all the energy necessary to maintain the loop was deposited near one chromosphere-transition region interface. Their calculations were, however, idealized. They joined two separate steady state solutions at a single point where the localized energy deposition took place, resulting in a discontinuity in the temperature gradient. In addition, their calculations did not include gravity.

Mariska (1988) used a time-dependent numerical model to examine flows driven by highly asymmetric heating. Figure 7.5 is an example of the results of a time-dependent numerical simulation with highly localized heating near one footpoint. The initial model is in hydrostatic equilibrium, with heat provided by a uniform volumetric energy deposition. Over a period of a few hundred seconds,

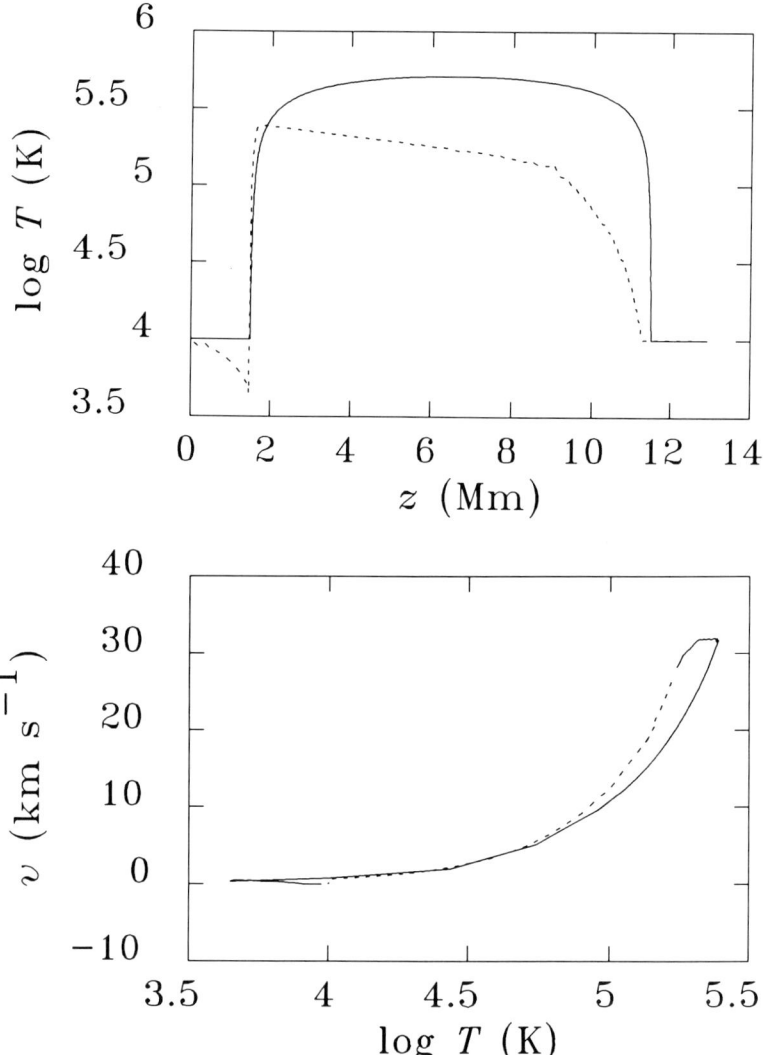

Fig. 7.5. The initial and final temperature structures and the final velocity structure for a numerical simulation of an asymmetrically heated loop. The solid line shows the initial temperature distribution, while the dashed line shows the final distribution. In the velocity plot, the solid line indicates material on the left side of the center of the loop and the dashed line indicates material on the right side (from Mariska, 1988).

all but a fraction of a per cent of this uniform background heating was removed and deposited in a band 100 km wide beginning 100 km above the base of the left transition region. After evolving for 2000 s, the model has the temperature distribution shown by the dashed curve. This highly asymmetric heating has produced a profound change in the temperature structure of the transition region and corona. The transition region on the left side is now steeper to conduct the excess energy toward the lower transition region where it can be more efficiently radiated away. The right transition region is now shallower in response to the reduced energy it must conduct to the lower transition region on the right.

The second panel in the figure shows the velocity as a function of temperature at the end of the simulation. The upflowing side of the loop (as viewed by an observer looking down from above) is plotted as a solid line and the downflowing side is plotted with a dashed line. Since the temperature peak occurs to the left of the center of the loop, the upflowing part of the curve has a "hook" in it near the temperature maximum. Both the upflowing and downflowing sides of the loop have velocities of about 10 km s^{-1} at 10^5 K, roughly what the observations exhibit. In this model, however, because the temperature gradient is much shallower on the downflowing side of the loop than on the upflowing side, all that would be observed at temperatures near 10^5 K would be the downflowing plasma. In this calculation the downflowing side is about a factor of 4 brighter than the upflowing side at the temperature of formation of the C IV resonance lines.

The highly asymmetric temperature profile of the final atmospheric model in this simulation has important observational consequences. Examination of the high-temperature end of the plot of velocity against temperature shows that there is a small range of temperatures near the maximum temperature for which an observer looking down on the loop would only see upflowing plasma. In any steady end-to-end flow with the temperature maximum located a substantial distance from the center of the loop, there must be some range of temperatures near the temperature maximum for which all the plasma in the range lies to one side of the center of the loop. Thus an observer looking down on the loop would only see blueshifted emission from lines formed in that temperature range. In the model shown in the figure, this range begins at about 2.2×10^5 K.

Mariska (1988) also found that the relative intensities of the upflowing and downflowing components of the flow change as a function of temperature. At low temperatures the downflowing side of the loop dominates the emission, leading to redshifted line profiles. At higher temperatures, the upflowing and downflowing components are roughly equal in intensity, leading to broadened line profiles with very little Doppler shift. Finally near the peak temperature in the loop, only blueshifted profiles are present. These contrast changes also result in changes in the width of the emission lines and may provide a partial explanation for the trend in the nonthermal broadening as a function of temperature.

While these flow models appear to provide a possible explanation for many observations in the transition region, they have one problem. In both the calculations of McClymont and Craig (1986, 1987) and Mariska (1988), the pressures in the model atmospheres were well below those observed in the quiet Sun. For a given loop size a higher pressure can be achieved by increasing the amount of heating. This, however, increases the peak temperature in the loop, which decreases the flow velocity at 10^5 K to velocities smaller than those observed. One solution to this dilemma is to place the flows in smaller loops. If the loops are made too small, however, they will violate the observational constraint that the emission in transition region lines must peak between 2 and 4 arc sec above the white-light limb.

This asymmetric heating model makes predictions that should help to distinguish it from other candidates for explaining the flows seen in the transition region. At high enough spatial resolution, it should be possible to distinguish the upflowing side of the loop from the downflowing side and see that the pattern is persistent in time rather than impulsive. Even at low spatial resolution, observation of blueshifted emission in high-temperature lines such as those from O VI or Ne VII in the quiet network would be strong evidence in support of this model. The small number of observations currently available suggests, however, that there are no sizable flows at the temperature of formation of Ne VII or Ne VIII (Hassler et al., 1991; Mariska and Dowdy, 1991).

The fact that these flows are driven by energy deposition that must be confined to a small region near one footpoint of a loop places constraints on any heating mechanism for the loop. It also suggests that both steady flows in loops and impulsive upflow followed by cooling and steady downflow may be related phenomena. If energy is deposited just above the chromosphere-transition region interface, a steady flow results. If on the other hand energy is deposited just below the interface, a spicule-like ejection into the loop might result. This would then be followed by cooling and downflow in the loop. Heating scenarios such as these are inherently time-dependent and require a more detailed solution of the full time-dependent set of mass, momentum, and energy equations.

7.5 Dynamic Transition Region Models

There are other possible explanations for the downflows seen in transition region lines. As we pointed out earlier, to within an order of magnitude, the amount of material injected into the corona by spicules equals the amount of material returning in the observed downflows. This match has led to a class of models that relate the spicule upflows and transition region downflows. At one extreme, one simply assumes that the spicule mass is injected into the corona, contributes to the total coronal mass, and eventually returns to the chromosphere (Pneuman and Kopp, 1977; 1978; Athay and Holzer, 1982). The complex magnetic structuring of the transition region and corona suggests that these phenomena must all take place within a single magnetic flux tube. This leads to a picture in which the upflow of plasma in spicules represents just one state in which an individual magnetic flux tube might exist (Athay, 1984). In this picture the spicule event represents a heating phase for a flux tube. After heating it can be in a quasistatic equilibrium or a cooling phase, which would lead to downflows.

This idea that the material in a magnetic flux tube need not be static has also led to transition region models based on the idea that the transition region and corona are heated episodically instead of continuously. Parker (1988, 1989) has argued that in the corona the heating should be localized and impulsive because magnetic field instabilities tend to release their energy over small areas and are impulsive. Sturrock *et al.* (1990) have applied this idea to the transition region. They suggest that microflares heat small elements of gas, which then cool radiatively, producing the observed UV and EUV transition region emission. Below 10^5 K, they argue that the observed emission measure implies low levels of heating. Each plasma element never reaches a temperature greater than about 10^5 K before cooling radiatively. Above 10^5 K, they argue that the observed emission measure implies that each element of gas is heated to coronal temperatures, followed by radiative cooling.

Raymond (1990) used this heating distribution to compute nonequilibrium models of the emission from a microflare heated atmosphere, finding reasonable agreement with the observed transition region line emission. While these calculations included all the elements in the Sturrock et al. (1990) heating model and the effects of departures from ionization balance on the cooling plasma, they did not include the interplay between different elements in the plasma produced by conduction and mass motions. Most discussions of these episodic models have focused on simple analyses of heating times and cooling times. At that level, the models appear to offer some promise of accounting for both the observed downflows and the emission measure distribution.

Mariska (1987) has investigated the consequences of one episodic heating model. He used numerical simulations to examine the response of a model transition region and corona to a reduction of heating that should lead to a cooler atmosphere and then a reheating of that cooled atmosphere. Earlier calculations (e.g., Peres et al., 1982; Mariska et al., 1982) showed that small changes in the atmospheric heating will generally result in only minor dynamical changes. Mariska therefore concentrated on large changes. Beginning with a quiet atmospheric model he linearly reduced the heating rate to 10% and 1% of the initial value over time scales of 100, 1000, and 2000 s. The atmosphere always evolved in roughly the same manner.

Figure 7.6 shows one example in which the heating rate was reduced to 1% of the initial value over a time scale of 100 s. This time is small compared to the radiative cooling time for the loop. When the heating is removed, the loop initially continues to cool in the manner it did while in a steady state. As the energy balance shown in Figure 7.2 shows, this means that the top portion of the loop will cool by conduction, while the lower regions cool by radiation. The initial cooling reduces the pressure, resulting in a

Dynamic Transition Region Models 221

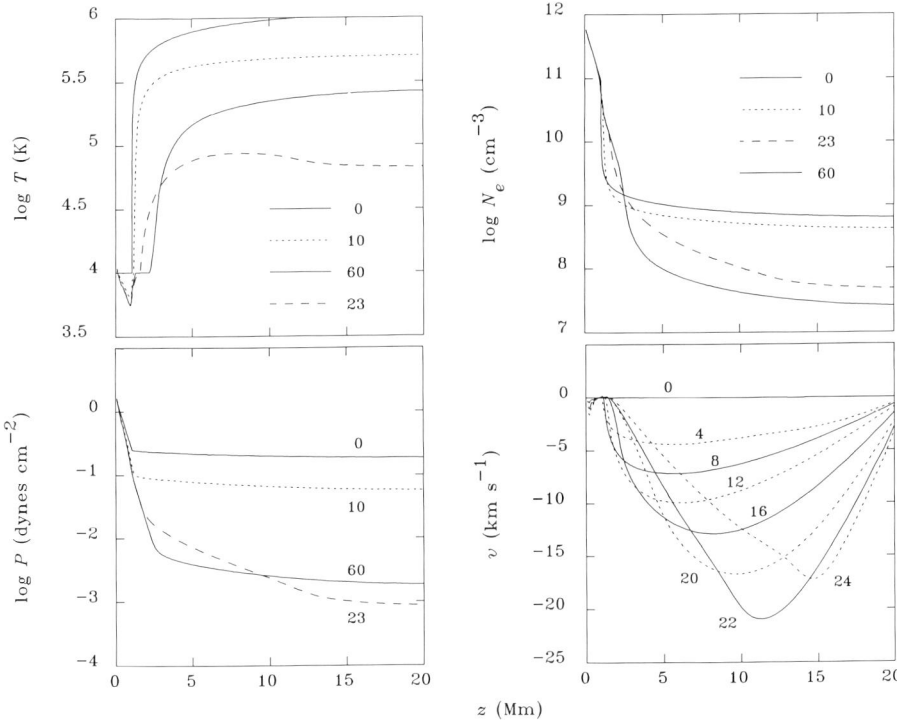

Fig. 7.6. The response of the temperature, pressure, electron density, and velocity to a reduction over a time scale of 100 s of the heating rate to 1% of the value required to maintain the initial atmospheric model. The times that label the curves are in units of 100 s (from Mariska, 1987).

force imbalance that produces downflows as the atmosphere reduces the amount of material it supports.

As the atmosphere continues to cool, the downward velocities increase and they begin to carry significant quantities of energy out of the coronal portions of the loop. These downward mass flows quickly overwhelm conduction as an energy loss mechanism in the coronal portions of the loop. In the lower portions of the loop, the decelerating plasma acts as a heat source.

As the cooling progresses, the changes in density and pressure at each point in the atmosphere begin to accelerate. The flow speed approaches its peak value and the coronal densities change more rapidly. Finally the atmospheric temperature distribution

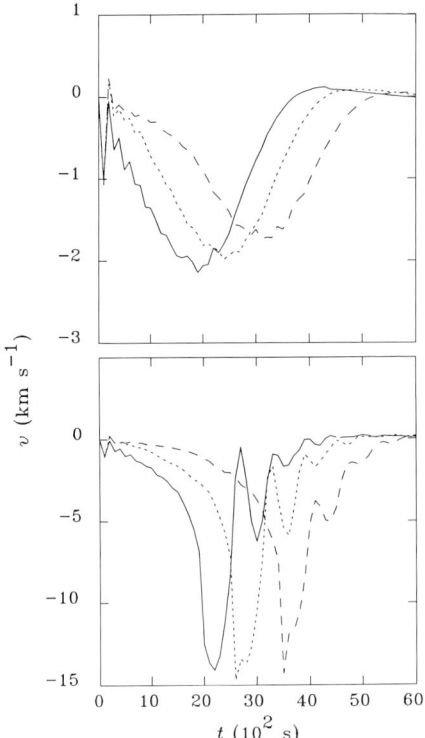

Fig. 7.7. The time evolution of the velocity at 10^5 K for cooling calculations in which 90% (top panel) and 99% (bottom panel) of the energy required to maintain an initial atmospheric model has been removed. The solid lines are for a time scale for the heating reduction of 100 s, the short dashed lines are for a time scale of 1000 s, and the long dashed lines are for a time scale of 2000 s (from Mariska, 1987).

overshoots its final equilibrium value and slowly increases back to a stable value.

Clearly, this cooling process produces significant downflows and abundant amounts of cool material. The key question is, do these downflows match the observations? Mariska (1987) found that as the plasma cooled the slope of the emission measure at low temperatures changed very little. Thus the cooling plasma does not account for the steep negative slope at low temperatures.

Figure 7.7 shows how the velocity at 10^5 K behaves as a function of time for a series of cooling calculations. When the heating is reduced to 10% of the initial value, the velocities produced are small compared with those measured in the quiet Sun. When the heating is reduced to only 1% of the initial value, the velocities are much larger. They exist, however, for only a limited time, roughly 500 s in these cases. Moreover, the peak velocities occur when the 10^5 K region is undergoing rapid cooling. During the 500 s that the large downflows are present, Mariska found that the emission measure at 10^5 K dropped by more than an order of magnitude. Thus the peak downflows are present when it is increasingly difficult to observe the downflowing plasma. In addition, the short lifetimes of the downflowing plasma require many structures in the field of view with only some of them cooling at any given time. In this picture, however, the emission measure from the stationary plasma at 10^5 K would overwhelm the emission measure from the cooling plasma. Thus it appears that simple cooling scenarios for the transition region will not reproduce the observed behavior of the quiet transition region.

Mariska (1987) also found that heating a loop after it has cooled to a low-temperature, low-density equilibrium will not reproduce the observed behavior of spicules. Possible heating scenarios are, however, more varied than cooling scenarios. Mariska only looked at a restoration of the heating necessary to maintain the initial loop configuration. Athay (1984) on the other hand suggested a much larger heating rate for a briefer period. Based on Mariska's calculations, however, it appears that such a large heating rate applied uniformly in the loop would result in flare-like behavior. The low-density portions of the initial loop model would heat rapidly to coronal temperatures followed by the steady evaporation of heated material from the top of the chromosphere. A heating model in which this energy was deposited in the upper layers of the chromosphere may reproduce the observed behavior of spicules.

Several investigations have examined time-dependent numerical models in which heat or a force applied in the chromosphere produces spicule-like mass ejections into the transition region and corona. Early calculations, which did not include radiation and conduction (e.g., Hollweg, 1982; Suematsu et al., 1982; Sterling and Hollweg, 1988), suggested that spicule-like ejections could be produced. More recent calculations, which include these dissipative effects (e.g., Mariska and Hollweg, 1985; Sterling and Mariska, 1990; Sterling et al., 1991), however, suggest that radiation removes most of the energy deposited in the chromosphere, reducing the heights and velocities reached by the disturbances.

Common to all these calculations, however, is the fact that they show that disturbances in the chromosphere usually result in low-amplitude fluctuations in the velocity at transition region temperatures. Thus the natural velocity fluctuations associated with convection in the lower regions of the solar atmosphere may be the source of the ubiquitous low-amplitude fluctuations seen in the transition region.

7.6 Cool Loop Models

Studies of static solutions to the momentum and energy balance equations have also revealed the possibility of solutions in which conduction plays a minor role. Hood and Priest (1979) first noted that for a given loop length and heating rate the usual conduction dominated solution to the energy equation has the maximum allowed pressure. If the pressure were larger, there would be more radiation than the heating could sustain and the loop would cool to a new equilibrium at a much lower temperature and pressure.

7.6.1 Static Cool Loop Models

In the cool loop solutions to the static energy balance equation, conduction plays no role. Thus the static energy balance equation becomes

$$H = L, \qquad (7.35)$$

heating is balanced everywhere locally by radiation. There are many possible forms for the heating term. Here, for simplicity, we assume that H depends only on position. Also for simplicity, we follow McClymont and Canfield (1983) and assume that the temperature dependence of the radiative loss function below 10^5 K obeys the power law

$$P(T) = \Lambda_0 T^3, \qquad (7.36)$$

where $\Lambda_0 = 6.46 \times 10^{-37}$ erg cm^3 s^{-1} K^{-3} gives agreement with the radiation loss rates at 10^5 K listed in Table 2.4. Taking $n_p = n_e$, the energy balance equation then becomes

$$\Lambda_0 n_e^2 T^3 = H, \qquad (7.37)$$

showing that the temperature in this case is directly proportional to the heating rate. This means that in regions of uniform pressure, the temperature and heating profiles will have similar forms, and any local changes in the heating rate should produce local deviations in the temperature.

The static momentum equation is

$$\frac{dP}{dh} = -m_p n_e g, \tag{7.38}$$

where h is the height, not the position along the loop. Using the ideal gas law to eliminate n_e and our temperature equation, we have

$$\frac{dP}{dh} = -\frac{m_p g \Lambda_0 P^3}{8k^3 H}. \tag{7.39}$$

At any height h, the pressure is then given by

$$P(h) = P_0 \left[1 + \frac{2}{H_{T_0}} \int_{h_0}^{h} \frac{dh}{f(h)}\right]^{-1/2}, \tag{7.40}$$

where P_0 is the base pressure,

$$H_{T_0} = \frac{2kT_0}{m_p g} \tag{7.41}$$

is the gravitational scale height at the base, and

$$f(h) = \frac{H(h)}{H(0)} \tag{7.42}$$

is the heating function normalized to its base value.

Equation (7.40) shows that for the cool solutions the pressure at any height in the loop depends only on the heating at all levels below that height. It also depends on the base pressure, which is given by the base heating and base temperature. In hot loops, on the other hand, because of thermal conduction, the pressure at each height depends on the heating at all other locations in the loop.

Figure 7.8 shows the cool solution to the full static momentum and energy balance equations for a uniformly heated loop with a height of 1500 km (Klimchuk and Mariska, 1988). This cool loop solution was built on a model chromosphere with a fixed temperature of 20,000 K. At the top of the 20,000 K region the pressure is 0.2 dyn cm^{-2}, the accepted quiet-Sun value.

In hot loops the temperature rises abruptly in the transition region and then remains roughly constant in the corona. In cool loops

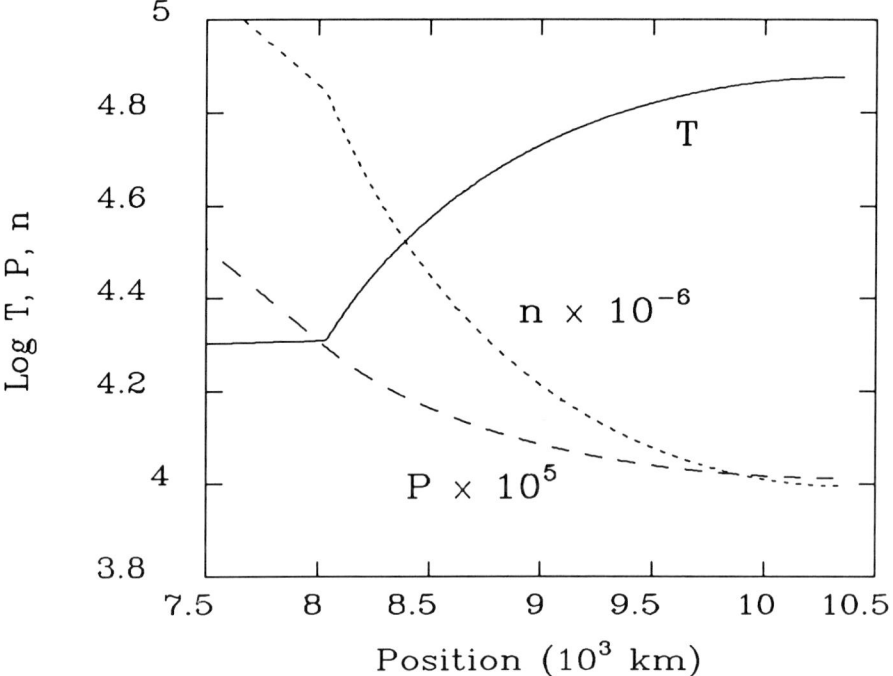

Fig. 7.8. The static equilibrium structure of a uniformly heated short loop of radius 1500 km. The temperature (K), total pressure (dyn cm^{-2}), and total number density (cm^{-3}) are plotted as functions of position along the loop axis. Only half the loop is shown. The loop apex is located at a position of 10,400 km and the base of the cool transition region is at 8000 km (from Klimchuk and Mariska, 1988).

the temperature increases gradually throughout the loop. This difference results from the fundamental difference in the energy balance between these two solutions. Thermal conduction plays the dominant role in the hot solutions by transporting energy from the corona to the transition region, with the steep temperature gradient necessary to maintain the large conductive flux. Local energy balance dominates the cool solutions.

For heating that is constant per unit volume, radiation also must be uniform per unit volume. This means that in the cool solutions the temperature must increase to offset the decrease in the density caused by gravity. If the heating increases with height, the temperature will rise more quickly; if the heating decreases with height the temperature will rise less quickly. The maximum temperature attainable by a cool loop with uniform energy deposition per unit volume is 10^5 K, the temperature at which the constant pressure form of the radiation loss curve we are using $P(T)/T^2$ has its first negative slope. Beyond this point a temperature increase cannot offset a density decrease, and the radiation must decrease. Energy balance is then impossible for the uniform heating case, and becomes more difficult even with heating that decreases with height, since $P(T)$ itself has a negative slope beyond 10^5 K.

This maximum temperature places an upper limit on the height of a cool loop. With heating that is uniform in space, the maximum loop height is 2000 km. For heating that is constant per unit mass, the maximum height is 8000 km (Antiochos and Noci, 1986). Even greater heights are possible for other choices of the radiation loss rate and the heating rate (Woods et al., 1990). In contrast, there is no limit to the height hot loops can reach.

Note that because of the role of gravitational stratification the limit for the cool loops is on the height, not the length. Thus long low-lying cool loops are possible. This important role for gravity also manifests itself in the pressure distribution in Figure 7.8. The gravitational scale height of a hot loop is generally large compared to its geometric height, resulting in a nearly constant pressure. In cool loops, on the other hand, the gravitational scale height is much smaller, resulting in the noticeable pressure stratification in Figure 7.8.

More complex heating terms do not alter the basic model we have outlined. Martens and Kuin (1982) examined cool solutions with a simple analytic model in which the heating rate was proportional to $P^\gamma T^\delta$ and the radiation followed a temperature behavior similar to the loss rates tabulated in Chapter 2. As in our simple model, the resulting energy equation contains only the pressure and temperature. For fixed values of γ and δ, Martens and Kuin found that stable solutions to the energy equation only exist for temperatures below 20,000 K. This lower peak temperature for the stable solutions than in our simple model is the result of the functional form of the radiative loss curve. If we use the actual Raymond (1979) loss rates tabulated in Chapter 2 instead of a T^3 law below 10^5 K, then the first negative slope in the constant pressure loss curve takes place at 20,000 K, the maximum in the constant pressure radiation loss curve.

From the maximum value of the temperature, an integration of the energy and hydrostatic equilibrium equations immediately determines the height of a cool loop. For reasonable choices of δ and γ, this height is roughly the hydrostatic scale height at the temperature maximum, about 1000 km, or less than 1.5 arc sec. This means that unless they emerge from the chromosphere at very shallow angles to the surface, the lengths of cool loops based on the Raymond (1979) radiation loss rates must be small. In addition it means that any changes in the magnetic field confining a cool loop that raised its height above 1000 km would result in the loop evolving to a hot conduction dominated state.

Antiochos and Noci (1986) further examined the nature of the cool solutions. As in our simple model, they ignored the peak in the radiation rates at 20,000 K and assumed that the first peak in the radiative losses occurs near 10^5 K. This then leads to a maximum loop height of about 5000 km, or a little less than 7 arc sec. While this value is still small relative to coronal size scales, it does satisfy the requirement imposed by the limb observations that the peak emission in transition region lines must take place 2–4 arc sec above the white light limb. Antiochos and Noci also examined the emission measure distribution produced by the cool loops. For a heating rate that was constant per particle, they found that the emission measure from the cool loops is proportional to T_m^{-3}, where T_m is the apex temperature in the loop. This dependence agrees with the observed emission measure slope. For constant heating per unit volume, however, the emission measure is proportional to T_m^{-1}. For reasonable parameterizations of the heating rate, Antiochos and Noci found that the observed emission measure could be produced only with a collection of loops with different peak temperatures.

7.6.2 Combined Hot and Cool Models

Because of the magnetic structuring of the atmosphere, most modelling efforts have centered on loops. In most models the entire loop has been assumed to contain either a cool solution or a hot solution to the momentum and energy balance equations. Observations, on the other hand, have not been good enough to prove that both types of loops exist. Instead, as our emission measure analysis showed in Chapter 6, they imply that at low temperatures radiation must be balanced by heating. Thus it is conceivable that hot solutions to the conservation equations in the upper transition region join continuously with cool solutions in the lower transition region.

Woods et al. (1990) have pursued this idea and performed a detailed examination of the energy balance of the lower transition region. They attempted to construct solutions to the static energy and momentum equations that would account for all the transition region emission from the chromosphere to the corona. This requires a transition from a conductively dominated model high in the transition region to a radiation and heating dominated model at lower temperatures.

They first constructed a simple energy balance model for the portion of the transition region above the emission measure minimum. To do this they specified the coronal temperature, conductive flux, and pressure and then integrated the energy and hydrostatic equilibrium equations downward until they reached the emission measure minimum. The energy balance equation in this portion of the model included only conduction and radiation. It therefore resulted in a model that did not fit the observed emission measure temperature dependence without the use of filling factors at temperatures below the coronal maximum in the emission measure.

Woods et al. (1990) then continued this integration downward in temperature until the slope b of the emission measure curve matched the value they were seeking for the lower transition region. Below that point they determined the temperature gradient, pressure, and heating rate using a procedure similar to that used by Jordan (1980). The results of these calculations showed that for temperatures below about 40,000 K heating primarily balances radiative losses. Above 40,000 K conduction dominates the energy balance.

Based on a study of simple models with the heating specified by one or two fluxes that were exponentially damped over specified height intervals, Woods et al. (1990) concluded that for models with negligible conduction in the lower transition region the variation of the emission measure as a function of temperature is very sensitive to the value of the heating rate. In addition they found that the existence of a peak in the constant pressure radiative loss rates due to Lyman α cooling leads to singularities in the solutions of

230   *Physical Transition Region Models*

the force and energy balance equations. These singularities then limit the range of acceptable solutions. Because of the Lyman α emission peak, their radiation and heating dominated portion of the loop is much smaller than the complete cool loop models developed by others. The nature of the radiative losses in the low transition region is thus of critical importance in the cool loop models.

7.6.3 Radiative Losses at Low Temperatures

All optically-thin radiative loss calculations show a local maximum at a temperature near 20,000 K due to Lyman α emission. For quiet-Sun conditions, however, below about 60,000 K, the atmosphere is not optically thin in some transitions. For example in the Vernazza *et al.* (1981) average quiet-Sun model, the line center optical thickness in the Lyman α line at a temperature of 24,000 K and an electron density of 2×10^{10} cm^{-3} is about 300. At these low temperatures, other abundant elements that are present in neutral or singly ionized form also contribute significant radiation (e.g., Avrett, 1985). Many of these lines are also potentially affected opacity.

To examine the importance of opacity at low temperatures, Mc-Clymont and Canfield (1983) computed an effective radiative loss function for the Vernazza *et al.* (1981) very bright network element model. At low temperatures this radiative loss function was reduced by about an order of magnitude from the Cox and Tucker (1969) value and about a factor of 20 from the McWhirter *et al.* (1975) radiative losses. The local maximum in the radiative losses was, however, still present. Based in part on this reduction in the importance of the Lyman α peak and in part by computational convenience, McClymont and Canfield suggested that below 10^5 K a simple T^3 functional form for the radiative losses could be used.

Athay (1986a) has argued that a T^3 functional form represents too large a correction. He computed mean radiation loss rates per unit volume in Lyman α for a range of solar densities and optical thicknesses. The results show that up to a line center optical thickness of about 100, the standard optically-thin coronal approximation is adequate. For the conditions in the Vernazza *et al.* (1981) model, the reduction was significantly smaller than that found by McClymont and Canfield. Athay pointed out that the key problem is a proper treatment of hydrogen ionization and that the same problem exists for other significant radiators in the low transition region. A proper treatment is necessarily model dependent and thus introduces additional difficulties in attempts to model the structure and energy balance of the lower transition region. Cally and Robb (1991) have also argued that low ionization states of silicon are more important radiators in the 20,000–10^5 K range than earlier calculations have assumed. They note that increased radiation from these ions would change the slope of the radiation loss function from 3 to 1. Variations in elemental abundances also could be a factor. Clearly, the existence of cool loops with reasonable sizes and peak temperatures to contribute to the emission measure below about 2×10^5 K is highly dependent on the radiative losses between 10^4 and 10^5 K. Ultimately high spatial resolution observations in optically thin UV emission lines formed in this temperature range may provide the answer.

7.6.4 Flows in Cool Loops

If cool loops are contributing most of the emission at temperatures near 10^5 K, then the Doppler shift observations show that they must exhibit quasisteady flows. Klimchuk and Mariska (1988) and McClymont (1989) have investigated the physics of those flows. Both studies used a power law for the radiative loss rates at temperatures below 10^5 K and thus began with loops that had peak temperatures near 10^5 K. Each study, however, used a different approach to examine the flows.

Klimchuk and Mariska (1988) used a time-dependent approach similar to that taken by Mariska (1988). They first established a cool loop model and then slowly modified the energy deposition to produce a flow. When the energy was localized in a small region near one footpoint of the loop, the temperature structure altered to produce a maximum in the temperature with a shape similar to the Gaussian shape of the energy deposition. This result is consistent with the heating balancing the radiation at each location. The new temperature distribution was accompanied by a flow from the heated side of the loop to the other side. In none of the simulations, however, did a substantial flow velocity develop.

Applying an exponentially decreasing heating rate, however, resulted in a different temperature and velocity structure. The peak temperature moved to the side of the loop with the smaller heating.

232 *Physical Transition Region Models*

This is because the heating is decreasing with decreasing height in the leg of the loop with the smaller heating, which requires a steeper temperature gradient to keep the radiation and heating in balance. This exponentially heated model also had higher velocities, near 20 km s^{-1} at the peak temperature. The emission measures at each temperature on the upflowing side and downflowing side of the loop were, however, dominated by the temperature gradient. Thus the upflowing side, with a shallower temperature gradient, had more emission at all temperatures except those near the peak. Thus at most temperatures upflows would dominate the emission in transition region lines.

McClymont (1989) also found that small changes in the heating rate induced flows in cool loops. For a gradually changing heating rate, the result is again a temperature peak on the downflowing side of the loop, with a shallow temperature gradient on the upflowing side and a steep one on the downflowing side. McClymont, however, claimed that spectral line synthesis showed that redshifts would dominate at the highest temperatures. Clearly this area needs further investigation before we fully understand the details of the flows and their observational consequences.

7.7 Additional Physics

Because of the difficulties all the models have in explaining the shape of the emission measure curve, especially at low temperatures, recent theoretical work has focused on the possibility that additional physical processes need to be incorporated into the energy equation.

7.7.1 Diffusion

As we pointed out in Chapter 6, the steep gradients in physical quantities in the lower transition region inferred from empirical models suggest that diffusion may be an important process in those layers. Fontenla *et al.* (1990, 1991) have constructed the first physical models for the lower transition region that take this process into account.

Including diffusion adds extra terms to the energy equation. For a plane-parallel geometry in the steady flow case, Fontenla *et al.* (1990) derived the energy equation

Additional Physics

$$\frac{d}{ds}\left[v\left(\frac{5}{2}P - E_H n_1\right) - \left(\frac{5}{2}kT + E_H\right)\frac{n_p n_1}{(n_p + n_1)}V_A + F_c\right]$$
$$= v\frac{dP}{ds} - L, \tag{7.43}$$

where E_H and n_1 are the ionization energy and number density of hydrogen atoms, V_A is the ambipolar diffusion velocity, and F_c is the conductive heat flux. The first term in the square brackets on the left side of equation (7.43) is the thermal energy carried by mass motions. It consists of the enthalpy flux and a second term accounting for ionization. This second term represents the energy released for example if atoms ionize at one point and recombine at a second point liberating the energy that was required to ionize them. The second term in the square brackets is the electron enthalpy flux that results from the flow at the ambipolar diffusion velocity V_A, again including ionization. The third term in the square brackets is the usual conductive heat flux.

Along with diffusion, equation (7.43) includes the effects of ionization. Even without diffusion, the enthalpy flux term in the energy equation should include ionization energy at temperatures below about 20,000 K. At higher temperatures the gas quickly becomes fully ionized. Ionization energy is, however, also intimately related to the diffusion term in the energy equation. The ambipolar diffusion velocity is positive, since atoms diffuse upward toward regions of lower concentration. Thus the second term is the enthalpy flux of electrons flowing downward because of diffusion, along with the ionization energy they have available to liberate on recombining. Since this energy equation is for a hydrogen plasma, the upward enthalpy flux of neutral hydrogen balances the downward enthalpy flux of protons, and those terms do not appear.

Fontenla et al. (1990, 1991) solved the static version of this energy equation with a lower boundary at the 7940 K level in model C of Vernazza et al. (1981) and assuming values for the flux represented by the left side of equation (7.43). They then used an iterative procedure to correct an initial model to one that satisfied the energy and hydrostatic equilibrium equations, including a detailed treatment of the statistical equilibrium and radiative losses associated with hydrogen and optically-thin radiative losses for the other species. They found that in the lower transition region downward

234 *Physical Transition Region Models*

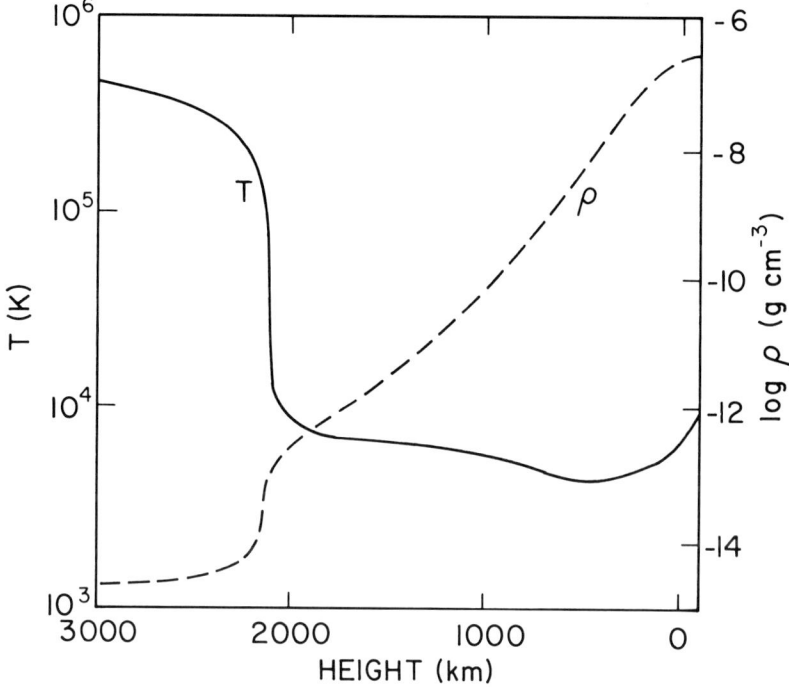

Fig. 7.9. The temperature and density in the solar atmosphere as functions of height for a model of the average quiet Sun that includes the effects of particle diffusion (courtesy E. H. Avrett, Center for Astrophysics).

energy flow was primarily due to a combination of the electron conductive heat flux and the hydrogen ionization energy due to the ambipolar diffusion. The ambipolar diffusion not only increased the downward energy flow at low temperatures, it also placed more neutral hydrogen at higher temperatures. Compared with the usual energy balance models, the net effect was to reduce the temperature gradient at temperatures below about 20,000 K and increase it above that level. In particular the 20,000 K plateau in the temperature distribution present in the Vernazza *et al.* (1981) models was no longer present. Instead the Lyman α line formed in the 40,000–70,000 K region.

Figure 7.9 shows how a model that includes diffusion alters the temperature distribution in the Vernazza *et al.* (1981) empirical models. This figure should be compared with Figure 1.7, which, for temperatures below 4.47×10^5 K is based on model C of Vernazza et al. (1981). Most striking in this figure is the absence of a temperature plateau at 20,000 K.

Avrett (1991) notes that in the 9000–20,000 K region the diffusion velocities are in the range from 0.1 to 2 km s^{-1}. These velocities are comparable to the measured downflow velocities seen in some lines in the lower transition region (e.g., Hassler et al., 1991). Thus the energy transport features of the model could equally well be explained by ordinary mass flows. Some effects produced by diffusion, for example the fact that it causes neutrals to move upward and ions to move downward, cannot, however, be produced by ordinary mass flows. In addition, both the diffusion velocities and the measured downflow velocities are small compared with the roughly 10 km s^{-1} turbulent velocities measured near 20,000 K (see, e.g., Figure 5.2). Of course the nature of the nonthermal broadening is unclear. The key point is that, while diffusion may be important in the lower transition region, a full understanding of that importance will require energy balance modelling that includes mass flows.

7.7.2 Turbulent Conduction

Fiedler and Cally (1990) and Cally (1990) have shown that it is possible to match the observed emission measure at low temperatures by invoking a nonclassical expression for the thermal conductivity. They argued that the observed nonthermal broadening signifies the presence of turbulence, leading to a turbulent conductivity of the form

$$\kappa_T = \phi \rho c_P u l, \tag{7.44}$$

where c_P is the specific heat at constant pressure, u is the rms turbulent velocity in one direction, l is the mixing length, and ϕ is a constant of order unity. For a mixing length with a value of 100 km at 20,000 K and a $T^{-\alpha}$ dependence, Cally (1990) found that $\alpha = 1.5$ resulted in predicted emission measures that were in excellent agreement with observations. Note that the turbulent conduction has a dramatically different temperature dependence than classical conduction. For constant pressure $\kappa_T \sim T^{-(0.5+\alpha)}$, so that $\alpha = 1.5$, gives $\kappa_T \sim T^{-2}$. This temperature dependence means that in the upper transition region classical conduction would dominate over turbulent conduction.

Physical Transition Region Models

Models based on turbulent conductivity have a completely different character than those we have discussed above. Below about 1.6×10^5 K, turbulent conduction quickly becomes orders of magnitude larger than classical conduction. This larger conductivity results in a shallower temperature gradient, resulting in the turnup in the emission measure at low temperatures. In fact the conductivity is so large that the energy balance in the model becomes conduction dominated just as it is in the upper transition region. All the radiative losses in the lower transition region remove only a small amount of the energy in the conduction term of the energy equation. Thus one consequence of this model is that the turbulent conduction dumps about 1.5×10^5 erg cm^{-2} s^{-1} into the upper chromosphere. This additional energy input would then have to be disposed of in the chromosphere.

7.7.3 Conduction Across the Magnetic Field

From the point of view of matching the observed emission measure distribution at low temperatures, turbulent conductivity has the desirable property that it increases with decreasing temperature. Conduction perpendicular to a magnetic field shares this property. The coefficient for thermal conductivity by protons perpendicular to the magnetic field is (Spitzer, 1962)

$$\kappa_\perp = 2 \times 10^{-16} \frac{n_p^2}{T^{1/2} B^2} \quad \text{erg s}^{-1} \text{ K}^{-1} \text{ cm}^{-1}. \tag{7.45}$$

Taking $n_p \approx n_e$ and assuming constant pressure, we have

$$\kappa_\perp \approx 2 \times 10^{-16} \frac{P_0^2}{T^{5/2} B^2}, \tag{7.46}$$

where $P_0 = n_e T$. Thus the conductivity perpendicular to the magnetic field has a temperature dependence near that of the turbulent conduction coefficient derived by Cally (1990).

The magnetic field, however, strongly inhibits perpendicular conduction. Using equation (7.23) and assuming constant pressure, we have

$$\frac{\kappa_\perp}{\kappa_\parallel} = 8 \times 10^{-12} \frac{P_0^2}{T^5 B^2}. \tag{7.47}$$

For a typical pressure of 10^{15} cm^{-3} K, this ratio is $8 \times 10^{-7} B^{-2}$ at 10^5 K and $0.08 B^{-2}$ at 10^4 K. Thus conduction perpendicular to the magnetic field does increase in importance at lower temperatures. It is still, however, much smaller than parallel conduction.

Increasing the area available for cross-field conduction can provide one way out of this dilemma. Rabin (1986) suggested that this was the case in the lower transition region. If cool features extend along the magnetic field lines into the hotter transition region, then a combination of parallel and perpendicular thermal conduction could explain the observed emission measure distribution if the surface area of the cool features is about a factor of 1000 greater than the surface area of the spherical Sun.

Athay (1990) has further quantified this idea by examining the consequences of one such geometry. He suggested that the solar transition region consists of a pattern of alternating cool peaks and hot valleys. For a given value of the downward conductive flux at the top of the atmosphere, the hotter regions, which are produced by the natural fluctuations in coronal heating, would have thicker transition regions than adjacent cooler regions. These height differences then provide locations where cross-field thermal conduction could operate. Using observed emission measures, Athay then concluded that the ratio of the height of the cool features to their width must be about 160. Taking for the height the roughly 2000 km width of the transition region at the limb, this yields a width of about 12 km.

While the model is appealing, it still does not explain why the usual parallel conduction does not remove all the energy. For it to operate something must inhibit conduction parallel to the magnetic field. Rabin and Moore (1984) avoided this problem by proposing that the lower transition region consists of cool loops threaded by finely divided strands of electric current directed along the loop. Energy generated by these current elements, which must be 1 cm–1 km in size, would then be transported to the surrounding medium by cross-field conduction, again leading to roughly the observed emission measure behavior at temperatures below about 2×10^5 K.

8

The Transition Region in Perspective

In this final chapter, we first briefly look at stellar transition regions. Then we discuss what we know and what we need to find out about the solar transition region.

8.1 Stellar Transition Regions

Emission in the cores of the Ca II H and K lines and in the He I 10830 Å line betrays the presence of a chromosphere in many cool stars other than the Sun. Only with the advent of rocket and satellite observations, however, has it been possible to study transition regions and coronae on other stars. While several satellite and rocket experiments detected evidence for material in the atmospheres of other cool stars at transition region temperatures (e.g., Evans *et al.*, 1975; Jamar *et al.*, 1976; Dupree, 1975; Vitz *et al.*, 1976), it was the *International Ultraviolet Explorer* (*IUE*) satellite that provided UV spectral data in sufficient quantity to explore fully this temperature range in cool stars. *IUE*, much like *Skylab*, marked a watershed in the study of cool star atmospheres.

In this section we briefly summarize the main results of the *IUE* observations of cool star transition regions. Our focus is primarily on single stars. There is, of course, now a vast literature on cool star chromospheres and transition regions. Excellent reviews have been provided by Linsky (1980) and Jordan and Linsky (1987).

8.1.1 Line Intensity Observations

IUE provided the first moderate spectral resolution survey of cool stars in the spectral range from 1200 to 2000 Å. For main

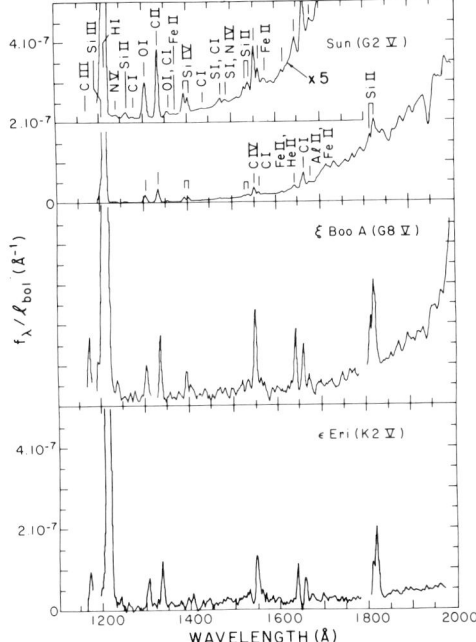

Fig. 8.1. Sample *IUE* spectra of G–K dwarfs. The solar spectrum at the top is a moderate-resolution rocket irradiance spectrum smoothed to the resolution of the low-dispersion mode of *IUE* (from Ayres et al., 1981).

sequence stars, low-resolution *IUE* observations show that stars from mid-F to mid-M have emission line spectra in this wavelength range that are similar to that of the Sun. Figure 8.1 shows a sample of low-resolution *IUE* spectra of main sequence stars in this wavelength range. Generally the same emission lines are present as in the solar spectrum shown at the top. In stars with spectral types later than the Sun, the emission lines are easier to detect because the photospheric continuum in this wavelength range decreases with decreasing effective temperature. By spectral type M, the continuum is essentially gone in this wavelength range. This problem of emission from the photospheric continuum overwhelming the transition region emission lines is reduced in solar observations by observing the transition region just above the white-light limb.

For giants and supergiants in this spectral range, transition region emission lines are not always present (e.g., Jordan and Linsky, 1987). Late giants show emission in the C IV resonance lines for spectral types from mid-F to late G. Spectral types later than K1 III show no C IV resonance line emission. Late supergiants show C IV line emission only in the early G spectral classes.

Later spectral type giants and supergiants still have a rich UV emission-line spectrum. The emission lines, however, tend to be fluorescent lines of O I, S I, Fe II, and CO. For example, the O I emission is not due to normal collisional excitation, but instead to excitation of O I by the H Lyman β line. The excited O I atom then cascades back down through the upper level of the resonance line. Similar processes produce many of the emission lines seen in these very cool atmospheres. Jordan and Judge (1984) provide an excellent discussion of the various radiative processes taking place in these late giant and supergiant stars. At low spectral resolution these additional lines complicate diagnosing the characteristics of transition regions in the coolest giants and supergiants.

Observations with *IUE* of the Mg II h and k lines at 2802.7 and 2795.5 Å show a short wavelength asymmetry indicative of mass outflow in many giants and supergiants that show evidence for transition region emission (e.g., Dupree and Reimers, 1987). Near the K1 III location in the H-R diagram, there is a change from solar type transition regions and hot coronae to the absence of a transition region and instead a massive cool wind indicated by Doppler shifts in the Mg II lines. This division in the H-R diagram between solar type transition regions and hot coronae and no transition region and a massive cool wind was at first thought to be quite sharp (e.g., Stencel and Mullan, 1980). Now it is clear that there is considerable overlap between the two regions (e.g., Dupree and Reimers, 1987).

Besides these two distinct classes of late type giants and supergiants, there is a third class know as hybrid stars. These show evidence for both transition region emission and for massive cool winds. Thus while all late type main sequence stars show evidence for transition regions like that seen on the Sun, the situation is far more complex for the giants and supergiants.

8.1.2 Physical Conditions and Empirical Models

All the spectral analysis tools we have discussed in this book are applicable to late type stars that exhibit transition region emission lines. As the surface gravity decreases, however, so does the transition region electron density. Thus some additional electron density diagnostics have been developed for these objects.

For main sequence stars and giants and supergiants with transition region emission lines, the emission measure analysis proceeds in exactly the manner we outlined earlier for the Sun. The only difference is that we are dealing with the flux from the entire stellar surface and must, in the absence of any information about filling factors, assume that transition region material uniformly covers the surface. Emission measure analyses based on *IUE* observations have been completed for many bright main sequence stars ranging from spectral type F5 through K2 (Brown and Jordan, 1981; Jordan et al., 1986; Fernández-Figueroa et al., 1981; de Castro et al., 1982; Jordan et al., 1987). Figure 8.2 shows a typical emission measure curve for a star of spectral type later than the Sun. The values indicated by the data points with horizontal bars are the result of a standard emission measure analysis like that we have outlined. The full lines are the result of assuming that all of the emission measure is produced at a single temperature and then computing a locus of points by varying the temperature. This procedure can prove to be useful for some lines that form over broad temperature ranges.

Figure 8.2 also shows the application of a standard solar type density diagnostic. The dashed curves in the figure show the locus of emission measure points for the 1892 Å density-sensitive intersystem line of Si III for different assumed electron densities. Selecting the curve that best fits the trend of the surrounding emission measures determines the electron density at the temperature of formation of the Si III line. This is essentially equivalent to the solar method of ratioing the emission in an intersystem line to the emission measure of an allowed line formed at nearly the same temperature.

In some cases densities in late type main sequence stars also can be determined from the ratio of the Si III 1892 Å line to the C III 1909 Å line. Often, however, the 1909 Å line is barely detectable or

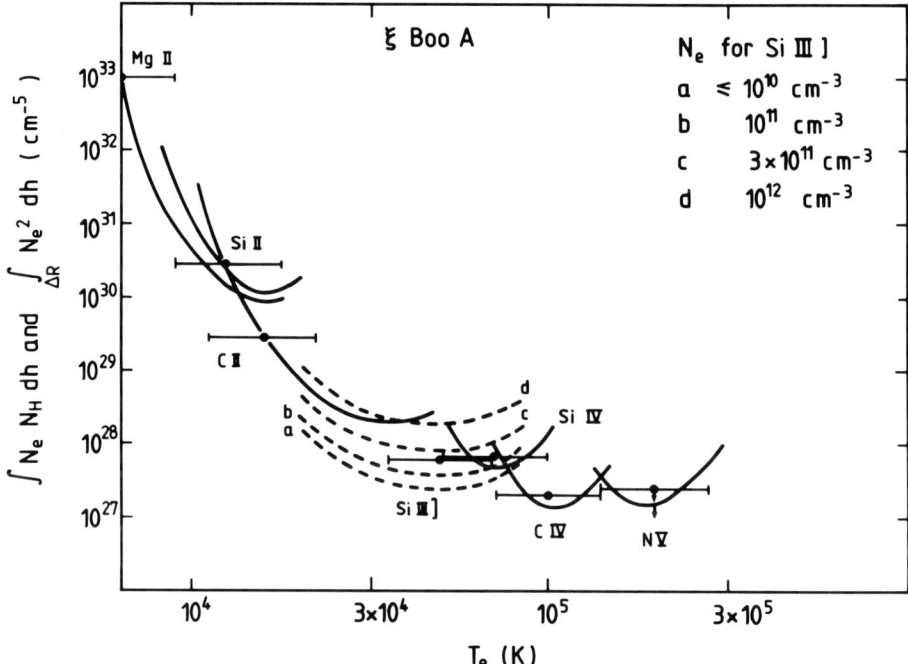

Fig. 8.2. Emission measure distribution for ξ Boo A. The loci of values of $\int n_e n_H \, dh$ (full lines) provide upper limits to the mean emission measure distribution. The values of $\int_R n_e^2 \, dh$ are the mean values required to give the observed flux, centered on the temperatures shown. The loci for Si III (dashed lines) are shown for densities in the range from 10^{10}–10^{12} cm^{-3} (from Jordan et al., 1987).

not visible at all, which yields a lower limit for the electron density at about 55,000 K.

As Figure 8.2 shows, the basic appearance of the emission measure in late type stars is quite similar to the solar emission measure in the same temperature range. The emission measure declines steeply with increasing temperature and shows some indication of flattening out near 10^5 K. Unfortunately, the wavelength range of *IUE* did not permit observations of lines with temperatures of formation in the upper transition region and corona. Thus the emission measure above about 2×10^5 K is difficult to determine.

X-ray observations provide some estimates of the emission measure in the corona for many stars modelled using *IUE* data. The *Einstein Observatory* Imaging Proportional Counter (Giacconi et *al.*, 1979) produced low-resolution X-ray spectra for many stars with hot coronae. For late type stars, the data consist of just flux measurements in a given X-ray band, or, in some cases, sufficient spectral information to estimate a coronal temperature. For example, for the ξ Boo A case shown in Figure 8.2, there are two X-ray flux measurements (Walter, 1981; Walter *et al.*, 1980) and only a rough temperature estimate of 10^7 K (Walter *et al.*, 1980). Any given X-ray detector has a known bandpass and sensitivity within that bandpass. When this is combined with calculations of the emissivity of a plasma containing a cosmic abundance of elements, the result is a curve of the emissivity in the detector as a function of plasma temperature. This is essentially a $G(T)$ curve for the detector. Thus an X-ray flux and a temperature estimate lead to an emission measure estimate in the corona and provide one high-temperature point on the emission measure plot. Assuming a uniform corona in hydrostatic equilibrium, these data also yield a crude estimate of the coronal density.

With the data outlined above, investigators have applied the same empirical model building techniques we outlined earlier to developing models and analyzing their energy balance. For the lower transition region, where there are many emission lines available with *IUE*, these emission measure based models of main sequence stars are remarkably similar to solar models. The electron pressures range from about 0.03 to 2 dyn cm^{-2} and, near 10^5 K, all of the stars have steep transition regions (Jordan and Linsky, 1987). At 10^5 K, the values of the inferred temperature gradient vary by roughly an order of magnitude on each side of the solar network value we derived in Chapter 6 (Jordan *et al.*, 1987). Because of the uncertainties associated with density determinations, however, these values are of course subject to large uncertainties.

Going from these empirical models of the lower portions of the transition region to a full empirical model that includes the corona is more difficult. Even when there is a coronal emission measure estimate, there is little or no data to help fill the gap between 10^5 K and the corona. Often the investigator simply assumes that the emission measure in that temperature range follows a $T^{3/2}$ relationship. The argument usually is made that this assignment is based on the behavior of the solar emission measure (e.g., Jordan *et al.*, 1987). As we showed in Chapter 4, however, there are considerable variations from a $T^{3/2}$ relationship for the solar emission measure. Thus, until there are sufficient measurements of stellar fluxes in emission lines formed in the upper transition region of late type stars, we should be cautious about placing much faith in the empirical models for that temperature range.

Energy balance analyses of empirical models for late type main sequence stars yield results that are remarkably similar to those derived from solar empirical models. Below 10^5 K thermal conduction is not important and the sizable radiative losses must be balanced by some form of local energy input (Jordan and Linsky, 1987). In those portions of the transition region where conduction is important to the energy balance, it is often a source term in the energy equation. Thus in the region near 10^5 K many empirical models require that the heating term be a sink for energy, just as we found for the Sun (Jordan et al., 1987). For some stars, such as ξ Boo A, the energy carried downward by conduction exceeds the radiation from the entire transition region and the chromosphere down to the level where the Mg II resonance lines form. This is primarily due to the high coronal temperature assigned to the star. Our complete lack of understanding of the magnetic geometry probably also contributes to these extremes in the inferred energy balance.

8.1.3 High-Resolution Spectroscopy

For a few of the brightest late type stars with solar-like transition regions, it has been possible to use the high spectral resolution capability of *IUE* to examine nonthermal broadening and Doppler shifts in transition region emission lines. Table 8.1 summarizes some of these measurements. All the stellar measurements are from Ayres et al. (1988), except those of ξ Boo A, which are from Jordan et al. (1987). For main-sequence stars with solar-like transition regions, the nonthermal broadening behaves much like it does on the Sun, increasing with increasing temperature of line formation. All the values listed for the stellar measurements are, however, larger than the solar measurements. This may be due in part to the fact that most of the lines used to construct Figure 5.2 were intersystem or forbidden transitions, which should have little or no broadening due to opacity. The stellar measurements on the other hand, are based on allowed transitions, such as the C IV resonance lines that may have some opacity. In addition, they

Table 8.1. *Stellar and solar nonthermal broadening and redshift measurements.*

Star	Type	ξ (km s^{-1})		$v_{\rm rel}$
		1.3×10^4 K	1×10^5 K	(km s^{-1})
χ^1 Ori	G0 V	17	34	8
α Cen A	G2 V	14	28	4
ξ Boo A	G8 V	15	38	
α Cen B	K1 V	14	15	6
ϵ Eri	K2 V	15	22	6
β Dra	G2 Ib-IIa	51	97	18
α Aur Ab	F9 III	40	98	13
λ And	G8 III-IV + ?	26	63	8
Sun Quiet	G2 V	10	20	4
Sun Active	G2 V			8

sample the entire stellar disk, so they include stellar rotation. At 1.3×10^4 K, the sound speed is roughly 18 km s^{-1}, while at 10^5 K it is 49 km s^{-1}. Thus some velocities for the giants and supergiants are in excess of the local sound speed.

Measuring Doppler shifts in UV transition region emission lines is at the limits of the *IUE* spacecraft's abilities. Ayres and his collaborators have, however, been able to make these measurements for several stars (Ayres et al., 1983; Ayres, 1984; Ayres et al., 1988). For all the stars they have measured, the Doppler shifts behave as they do on the Sun. The velocities in the table are the average redshift velocities determined from emission lines of Si IV and C IV. These velocities are measured relative to cooler lines formed in the stellar chromosphere. For comparison the table also includes the C IV solar measurements averaged over the solar disk under the assumption that the flows are vertical. It is clear from the table that stars like the Sun exhibit Doppler shifts similar to those seen in the solar transition region, while the lower surface gravity giants show even larger downflow velocities.

As in the solar case, both the nonthermal broadening and the presence of apparent downflows in other late type stars have energetic implications for models of the transition regions and coronae. If we interpret the observed nonthermal broadening as an indication of the presence of propagating acoustic waves, as we did in Chapter 5, the energy flux in the waves is not sufficient to both balance the local radiation in the transition region and provide sufficient energy to the corona to maintain the inferred downward conductive flux. If we interpret the nonthermal broadening as an indication of the presence of propagating MHD waves, then for reasonable estimates of the magnetic field strength, more than enough energy is available to account for both (Jordan et al., 1987).

These similarities to the Sun in the high spectral resolution observations warn us that any serious conclusions about atmospheric structure and energy balance for late type stars that are based on empirical models like those we developed in Chapter 6 should be viewed with caution. These models simply do not predict the actual observed solar transition region structure. Since many main sequence late type stars are similar to the Sun in every transition region feature that we can observe, they are also likely to have atmospheric structures that just as stubbornly refuse to conform to our simple models.

8.1.4 Cool Stars and the Sun

Surveys with *IUE* and other satellites have provided a wealth of low spectral resolution data on stellar transition regions. Combining this data with other information on these late type stars allows us to place the Sun in an evolutionary context with these objects. Ground-based observations show that chromospheric activity as indicated by the intensity of the emission cores of the Ca II H and K lines declines with age in main sequence stars (Wilson, 1963). Rotation rate determinations also show that for late type main sequence stars the rotation rate declines with age (e.g., Kraft, 1967; Soderblom, 1983). Simon et al. (1985) examined 31 F7 to G2 dwarfs observed with *IUE* and found that transition region emission as shown for example by emission in the resonance lines of C IV also declines with age.

The picture that has emerged to explain these correlations of activity with age centers on the dynamo model for rotating, convective stars. In the dynamo model, interaction of magnetic fields with the stellar rotation and convection determines the magnetic field structure at the surface (e.g., Gilman, 1983; Glatzmaier, 1985). Stars with higher rotation rates should have stronger dynamos and thus more surface magnetic fields, which manifest themselves as increased chromospheric and transition region emission. Since the strength of the dynamo also depends on the depth of the convection zone and the convective turnover time, dynamo theory also predicts that surface magnetic fields should be stronger for stars of later spectral type, as is observed. Noyes *et al.* (1984) have unified these correlations with the discovery that chromospheric activity correlates well with the ratio of the rotation period to the convective turnover time, the Rossby parameter $P_{\rm rot}/\tau_{\rm c}$.

Simon *et al.* (1985) have extended this correlation to transition region emission and find that the emission falls off with increasing $P_{\rm rot}/\tau_{\rm c}$. The higher the temperature of formation of the emission, the more rapid the falloff. The correlation with age of these parameters is thought to be due to braking of the stellar rotation through the action of a stellar wind, which is coupled to the star through the magnetic field (e.g., Kraft, 1967).

While this picture is appealing, many details are yet to be fully understood. In particular the details of how the magnetic fields generated by the dynamo produce chromospheric, transition region, and coronal temperature plasmas and a stellar wind are unclear. Here the Sun is crucial. Only in the solar case can we make the detailed helioseismology observations that will aid in more fully understanding the subsurface layers that are critical to unraveling the nature of the dynamo. Only in the solar case do we have the spatial and spectral resolution to examine in detail the structure of the outer layers of the atmosphere to test theoretical models for the heating. Thus, while observations of late type main sequence stars help to place the Sun in its evolutionary context, it is our understanding of the Sun itself that will validate our picture of the relationship between stellar activity and other stellar parameters. Analyses of *IUE* observations of late type stars show that they are similar to the Sun. If we understand the detailed structure and energy balance of these layers in the Sun, then we understand the same layers in late type main sequence stars. The question then is, how well do we understand the outer layers of the solar atmosphere?

8.2 Physical Models and Observed Structure

We know a great deal about the solar transition region. At 5 arc sec spatial resolution, UV and EUV observations provide a complete picture of the distribution of emitting material on the solar surface and as a function of temperature. At higher spectral and spatial resolution the available data are less complete, especially in temperature coverage. From these data, however, we can distill several observed features of the transition region that any successful physical model must explain.

8.2.1 Observed Features

First, any successful model must reproduce the observed transition region density. We have excellent diagnostic line ratios in the low transition region, which give a consistent measurement of the electron density at temperatures near 60,000 K. Accurate measurements at higher temperatures are not available, hampering our ability to understand the structure of the upper transition region. To the accuracy of our current determinations, the measured densities at the upper and lower boundaries of the transition region also agree with accepted values in the chromosphere and corona.

The second observational feature that must be explained is the gross morphology of the transition region. This includes the separation into a cell-network pattern on the disk and the extended emission above the limb. These two features, along with the density measurements, produce the small filling factors.

While emission measure determinations suffer from many uncertainties, the general features we see in all determinations must be explained by any successful theory. Those key features are the negative slope at low temperatures, the minimum near 2×10^5 K, and the positive slope at higher temperatures. The actual values of those slopes vary from region to region (see, e.g., Figure 4.1), so we should be wary of theories that try to force a magic number like 3/2 on the slope in the upper transition region. We also should not ignore out of hand the possibility that some features in the emission measure distribution may be due to our use of invalid assumptions in the determination. This may be especially true at low temperatures.

Like the emission measure, the variation of the nonthermal broadening as a function of temperature is a general property of the quiet solar transition region. When we consider only optically-thin lines, the functional form shown in Figure 5.2 always emerges. Here we are in need of more data. There are large gaps in our knowledge of the behavior of the nonthermal broadening at temperatures above about 3×10^5 K. While there are a few data points there, they do not allow us to draw any firm conclusions about the behavior of the nonthermal broadening.

We do know that the nonthermal broadening as a function of temperature does change in different solar features. In some prominence spectra, the line widths are down to the thermal width. Thus one key to unraveling the meaning of the nonthermal broadening may be high spatial resolution studies that relate the variations in the nonthermal velocity throughout the temperature range of the transition region to changes in the magnetic field strength or inferred geometry.

Finally, we know that Doppler shifts are ubiquitous in the quiet solar transition region. In most of the transition region, we see persistent redshifts, while blueshifts are more impulsive. The characteristics of the redshifted emission are well known for emission lines formed at temperatures at or below about 10^5 K, with few observations available at higher temperatures. Thus we have not fully characterized the flow field throughout the quiet solar transition region.

8.2.2 Unknowns

Unfortunately, there is also much that we do not know about the transition region. The geometry is most important. This is the key ingredient for using the emission measure to deduce the temperature and density structure of the atmosphere. We have excellent density diagnostics, but little useful information on how to interpret the dV in the emission measure integral. The emission measure itself provides no information about the spatial arrangement of material along the line of sight from our instrument to the Sun. It only reveals how much material in that line of sight is contained in each temperature interval. Without guidance from other observations, we are free to rearrange the spatial distribution of transition region material along the line of sight in any manner we choose, provided that we put the right amount in each temperature interval.

Of course this lack of geometric information really represents a lack of understanding of the magnetic structure of the transition region. As we show in Chapter 7, to first order the magnetic field serves to channel the plasma to follow the field lines. We have direct observations that show us the distribution of magnetic field at the photospheric level and indirect observations that show the presence of loop-like structures in the corona, but do not yet know how the two are connected.

The large-scale structures we see in soft X-ray observations span appreciable fractions of a solar radius, often interconnecting active regions. These structures thus probably overlie smaller X-ray emitting features with sizes closer to the 35,000 km characteristic size of a supergranule cell. In addition, the features we see in soft X-ray pictures are primarily those with the highest emission measure. As we showed in Chapter 7, however, the magnetic field must expand to fill all the coronal volume. How these less visible coronal regions connect to the lower transition region remains uncertain.

Of equal importance to the question of the geometry and probably related to it is the question of how the transition region is heated. This question lies at the heart of most transition region modelling efforts. One of the few successes in transition region studies since *Skylab* has been the elimination of acoustic waves as a heat source for both the transition region and corona. The nature of the actual heat source, however, remains elusive. This question is of course intimately related to the coronal heating question, since many geometries provide most of the transition region heating by thermal conduction from the corona. Without acoustic waves as a heat source, magnetic field related heating has dominated theoretical thinking. Because of the connection of the transition region to both the chromosphere and the corona and the overwhelmingly different heating requirements of those two parts of the solar atmosphere, we should be prepared for the possibility that multiple energy sources are at work in the transition region.

8.2.3 Model Features

Given what we know and the gaps in that knowledge, how well do the empirical and theoretical models reproduce the transition region observations? Since the empirical models are derived directly from the emission measure distribution and density measurements, they reproduce those precisely. In their simplest form, however, they fail to reproduce the observed extended limb emission or the resulting small filling factors. Their requirement that at some temperatures the heating term be a sink rather than a source of energy also suggests that they are conceptually incomplete. This may in large part be due to the small number of assumptions about the geometry and the energy balance we can allow and still obtain a useful inversion of the emission measure. Until we understand better the transition region geometry, emission measure based empirical models probably will not improve.

Purely theoretical transition region models have more freedom. While they must ultimately satisfy the observational constraints, they do not have to be formulated in a way that lends itself to direct inversion of the emission measure distribution. All the models based on the conventional static energy balance equation containing only heating, conduction, and radiation in a simple spherically-symmetric geometry fail to reproduce the observed emission measure, and, of course, fail to reproduce the observed downflows. Models constructed in magnetic field geometries that diverge with height come closer to fitting the temperature region above about 2×10^5 K, but also fail at lower temperatures.

Only the newer models based on additional terms in the energy equation, such as ambipolar diffusion and thermal conduction perpendicular to the magnetic field or relying on the cool solution to the energy equation, show any promise of reproducing the observed emission measure. The diffusion-based models also offer some promise of removing the artificial temperature plateaus that have plagued chromospheric models. When flows are introduced, however, the small diffusion velocities are likely to be overwhelmed. In addition, the first efforts to introduce diffusion by elements other than hydrogen seem to suggest that these models may introduce unacceptable abundance separations.

Models that introduce thermal conductivity perpendicular to the magnetic field or that rely on cool solutions to the energy equation offer some hope of meeting all the observational constraints. In particular, because they are based on geometries that are not spherically symmetric and do not necessarily have to have all temperatures from the base of the transition region to the corona in a single structure, they may be able to reproduce the extended emission observed at the limb and the small inferred filling factors.

8.2.4 Transition Region Structure

What does the combination of the observations, empirical models, and physical models suggest about the structure of the transition region and its role in the energy balance of the outer layers of the solar atmosphere? Small filling factors combined with extended emission at the limb and persistent Doppler shifts argue that much of the transition region must be in structures that are not describable using simple static energy balance atmospheres. On the other hand, those same energy balance models, when modified for a diverging geometry appear to provide an adequate description of the transition region at temperatures above about 2×10^5 K. Thus the observations and models strongly suggest that the transition region consists of multiple structural components.

The rise in the emission measure at low temperatures may be a sign that the dominant structural element is changing at temperatures near 2×10^5 K. This also would be consistent with the apparent disappearance of downflows at temperatures above 2×10^5 K. Even the nonthermal broadening may be beginning to turn over at that temperature, though the observational data are sparse.

Figure 8.3 shows one possible view of the structure of the transition region. Here only a fraction of the transition region connects to the hotter corona. Much of it resides in smaller loops of various sizes and reaches various temperatures. The closed loops have size scales roughly the size of an individual patch of network emission ($\sim$10 arc sec) or smaller. They might consist of a mixture of structures, some satisfying the hot solution to the energy equation and some satisfying the cool solution. Thus the observed emission measure distribution might not come from a single type of magnetic structure. Instead, it may be a blending of emission from several physically distinct structures. This would account for the failure of models that assume the entire atmosphere is embedded in a single magnetic structure.

This picture of a multicomponent transition region with only some components reaching coronal temperatures means that only some fraction of the transition region emission can be powered by energy conducted inward from the corona. Thus conductive energy

Physical Models and Observed Structure

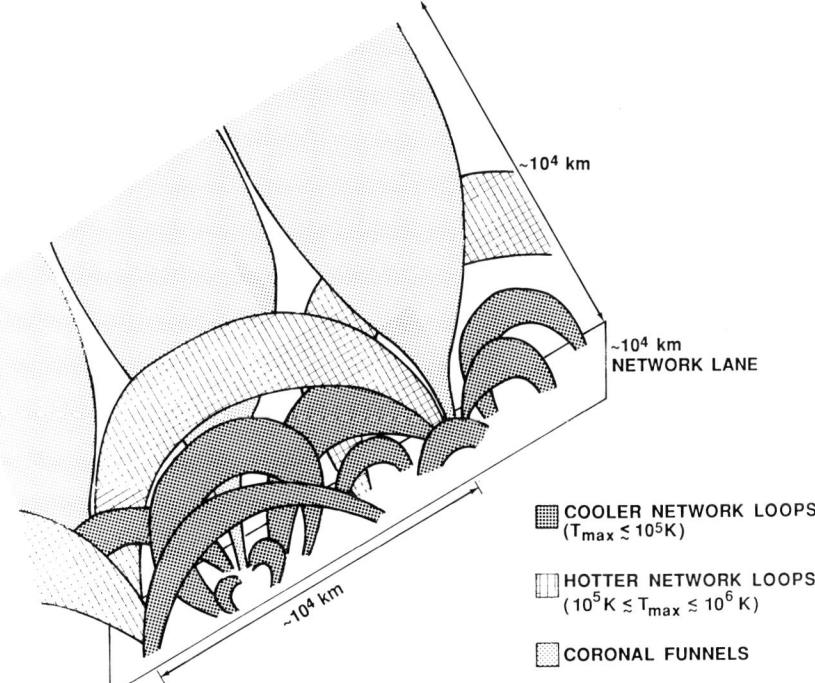

Fig. 8.3. One possible conceptual model for the three-dimensional geometry of the transition region (based on Dowdy et al., 1986).

loss estimates from the corona may need to be rethought. In addition, the total observed radiative losses of the transition region no longer need to be satisfied by the coronal heating mechanism. Some of those losses would be supported by whatever heating mechanism powers the smaller loops that fail to reach coronal temperatures. This may be the same process that heats the corona, or something different.

8.2.5 Future Work

What then must we do to make further progress toward understanding the transition region? In the area of observations, the most important requirements are high spatial and spectral resolution observations in the upper transition region. These would fill the gaps in our understanding of the flow field and nonthermal broadening as a function of position and provide more information on the transition region geometry. A full picture of the transition region geometry, the most significant unknown, will not emerge, however, until we have simultaneous imaging in UV and EUV lines combined with high spatial resolution visible images and magnetograms. Flying a magnetograph on the same spacecraft as UV and EUV instruments would of course be ideal.

Much theoretical work also needs to be done. On the practical side, density diagnostics for the upper portion of the transition region need further improvement, especially if new observations in the EUV portions of the spectrum become available. All the new ideas regarding the structure and energy balance of the transition region are based on physics that has not been fully explored in the transition region environment. Thus more work needs to be done to understand the possible roles of diffusion, turbulent conduction, and conduction perpendicular to the magnetic field.

A clear picture of the radiative properties of the lower transition region is also critical to many of the new ideas on transition region structure. Here the possible presence of abundance variations could cause important changes in the functional form of the radiative loss curve. More important, however, is a better understanding of the transport of radiation in the $10,000$–10^5 K range. Most of the cool loop solutions are only viable if the 20,000 K peak in the radiative loss curve due to Lyman α is not present. Currently there is no valid justification for removing it. More work needs to be done on how radiative losses behave in the lower transition region under various assumptions about the geometry and heating.

We have come a long way in our understanding of the solar transition region since the first rocket observations less than 50 years ago. The study of solar and stellar transition regions is young and still full of unanswered questions. New rocket and satellite instruments show great promise for answering many of those questions. And once we understand the solar transition region, we should be able to understand the outer layers of most of the stars on the lower main sequence. Above all we should examine the new observations without strong preconceived biases and be prepared for marvelous surprises.

When all has been said, the adventure of the sun is the great natural drama by which we live, and not to have joy in it and awe of it, not to share in it, is to close a dull door on nature's sustaining and poetic spirit.*

---

* From *The Outermost House* by Henry Beston. Copyright 1928, 1949, © 1956 by Henry Beston. Copyright © 1977 by Elizabeth Beston. Reprinted by permission of Henry Holt and Company, Inc.

REFERENCES

Aggarwal, K. M. (1984). *Solar Phys.*, **94**, 75.
Allen, C. W. (1973). *Astrophysical Quantities*, 3rd edn, Athlone Press, London.
Allen, J. W. and Dupree, A. K. (1969). *Astrophys. J.*, **155**, 27.
Aller, L. H., Ufford, C. W. and Van Vleck, J. H. (1949). *Astrophys. J.*, **109**, 42.
Almleaky, Y. M., Brown, J. C. and Sweet, P. A. (1989). *Astron. Astrophys.*, **224**, 328.
Anders, E. and Grevesse, N. (1989). *Geochim. Cosmochim. Acta*, **53**, 197.
Ando, H. and Osaki, Y. (1975). *Publ. Astron. Soc. Japan*, **27**, 581.
Antiochos, S. K. (1984). *Astrophys. J.*, **280**, 416.
Antiochos, S. K. and Noci, G. (1986). *Astrophys. J.*, **301**, 440.
Arnaud, M. and Rothenflug, R. (1985). *Astron. Astrophys. Suppl. Ser.*, **60**, 425.
Athay, R. G. (1966). *Astrophys. J.*, **145**, 784.
Athay, R. G. (1971). In *Physics of the Solar Corona* (Macris, C. J., ed.), p. 36, Reidel, Dordrecht.
Athay, R. G. (1976). *The Solar Chromosphere and Corona: Quiet Sun*, Reidel, Dordrecht.
Athay, R. G. (1981). *Astrophys. J.*, **249**, 340.
Athay, R. G. (1982). *Astrophys. J.*, **263**, 982.
Athay, R. G. (1984). *Astrophys. J.*, **287**, 412.
Athay, R. G. (1986a). *Astrophys. J.*, **308**, 975.
Athay, R. G. (1986b). In *Physics of the Sun*, Vol II (Sturrock, P. A., ed.), p. 51, Reidel, Dordrecht.
Athay, R. G. (1990). *Astrophys. J.*, **362**, 364.
Athay, R. G. and Dere, K. P. (1989). *Astrophys. J.*, **346**, 514.
Athay, R. G., Gurman, J. B., Henze, W. and Shine, R. A. (1983). *Astrophys. J.*, **265**, 519.
Athay, R. G. and Holzer, T. E. (1982). *Astrophys. J.*, **255**, 743.
Athay, R. G. and Thomas, R. N. (1956). *Astrophys. J.*, **123**, 309.
Athay, R. G. and White, O. R. (1978). *Astrophys. J.*, **226**, 1135.
Athay, R. G. and White, O. R. (1979a). *Astrophys. J.*, **229**, 1147.
Athay, R. G. and White, O. R. (1979b). *Astrophys. J. Suppl.*, **39**, 333.
Athay, R. G. and White, O. R. (1980). *Ap. J.*, **240**, 306.
Avrett, E. H. (1985). In *Chromospheric Diagnostics and Modelling* (Lites, B. W., ed.), p. 67, National Solar Observatory, Sunspot.

References

Avrett, E. H. (1991). In *Mechanisms of Chromospheric and Coronal Heating* (Ulmschneider, P., Priest, E. R. and Rosner, R., eds.), p. 97, Springer-Verlag, Berlin.

Ayres, T. R. (1984). *Astrophys. J.*, **284**, 784.

Ayres, T. R., Jensen, E. and Engvold, O. (1988). *Astrophys. J. Suppl.*, **66**, 51.

Ayres, T. R., Marstad, N. C. and Linsky, J. L. (1981). *Astrophys. J.*, **247**, 545.

Ayres, T. R., Stencel, R. E., Linsky, J. L., Simon, T., Jordan, C., Brown, A. and Engvold, O. (1983). *Astrophys. J.*, **274**, 801.

Bartoe, J. -D. F. and Brueckner, G. E. (1975). *J. Opt. Soc. Am.*, **62**, 949.

Bartoe, J. -D. F., Brueckner, G. E., Purcell, J. D. and Tousey, R. (1977). *Appl. Optics*, **16**, 879.

Baum, W. A., Johnson, F. S., Oberly, J. J., Rockwood, C. C., Strain, C. V. and Tousey, R. (1946). *Phys. Rev.*, **70**, 781.

Beckers, J. M. (1972). *Ann. Rev. Astron. Astrophys.*, **10**, 73.

Bely, O. and Faucher, P. (1970). *Astron. Astrophys.*, **6**, 88.

Berrington, K. A., Burke, P. G., Dufton, P. L. and Kingston, A. E. (1977). *J. Phys. B*, **10**, 1465.

Berrington, K. A., Burke, P. G., Dufton, P. L. and Kingston, A. E. (1985). *Atomic Data Nucl. Tables*, **33**, 195.

Bhadra, K. and Henry, R. J. W. (1980). *Astrophys. J.*, **240**, 368.

Bhatia, A. K., Doschek, G. A. and Feldman, U. (1980). *Astron. Astrophys.*, **86**, 32.

Biermann, L. (1946). *Naturwiss.*, **33**, 118.

Blaha, M. (1969). *Astrophys. J.*, **157**, 473.

Bohlin, J. D., Vogel, S. N., Purcell, J. D., Sheeley, N. R., Jr., Tousey, R. and Van Hoosier, M. E. (1975). *Astrophys. J. Lett.*, **197**, L133.

Boland, B. C., Dyer, E. P., Firth, J. G., Gabriel, A. H., Jones, B. B., Jordan, C., McWhirter, R. W. P., Monk, P. and Turner, R. F. (1975). *Mon. Not. Roy. Astr. Soc.*, **171**, 697.

Bonnet, R. M., Bruner, E. C., Jr., Acton, L. W., Brown, W. A. and Decaudin, M. (1980). *Astrophys. J. Lett.*, **237**, L47.

Bonnet, R. M., Lemaire, P., Vial, J. C., Artzner, G., Gouttebroze, P., Jouchoux, A., Leibacher, J. W., Skumanich, A. and Vidal-Madjar, A. (1978). *Astrophys. J.*, **221**, 1032.

Book, D. L. (1983). *NRL Plasma Formulary*, Naval Research Laboratory, Washington.

Boris, J. P. and Mariska, J. T. (1982). *Astrophys J. Lett.*, **258**, L49.

Borrini, G. and Noci, G. (1982). *Solar Phys.*, **77**, 153.

Bowen, I. S. and Edlén, B. (1939). *Nature*, **143**, 374.

Boyd, T. J. M. and Sanderson, J. J. (1969). *Plasma Dynamics*, Barnes and Noble, Inc., New York.

Braginskii, S. I. (1965). In *Reviews of Plasma Physics*, Vol I (Leontovich, M. A., ed.), p. 205, Consultants Bureau, New York.

Bray, R. J. and Loughhead, R. E. (1974). *The Solar Chromosphere*, Chap. 3, Chapman and Hall, London.

Breneman, H. H. and Stone, E. C. (1985). *Astrophys. J. Lett.*, **299**, L57.

Brown, A. and Jordan, C. (1981). *Mon. Not. Roy. Astr. Soc.*, **196**, 757.

Brueckner, G. E. and Bartoe, J. -D. F. (1974). *Solar Phys.*, **38**, 133.

Brueckner, G. E. and Bartoe, J. -D. F. (1983). *Astrophys. J.*, **272**, 329.
Brueckner, G. E., Bartoe, J. -D. F., Cook, J. W., Dere, K. P. and Socker, D. G. (1986). *Adv. Space. Res.*, **6**, No. 8, 263.
Brueckner, G. E., Bartoe, J. -D. F., Cook, J. W., Dere, K. P., Socker, D., Kurokawa, H. and McCabe, M. (1988). *Astrophys. J.*, **335**, 986.
Brueckner, G. E. and Moe, O. K. (1972). *Space Res.*, **XII**, 1595.
Brueckner, G. E. and Nicolas, K. R. (1973). *Solar Phys.*, **29**, 301.
Bruner, E. C., Jr. (1978). *Astrophys. J.*, **226**, 1140.
Bruner, E. C., Jr. (1981). *Astrophys. J.*, **247**, 317.
Bruner, E. C., Jr. and McWhirter, R. W. P. (1979). *Astrophys. J.*, **231**, 557.
Bruner, M. E. and McWhirter, R. W. P. (1988). *Astrophys. J.*, **326**, 1002.
Burton, W. M., Jordan, C., Ridgeley, A. and Wilson, R. (1973). *Astron. Astrophys.*, **27**, 101.
Byerley, A., McWhirter, R. W. P. and Wilson, R. (1978). *J. Phys. B*, **11**, 613.
Cally, P. S. (1990). *Astrophys. J.*, **355**, 693.
Cally, P. S. and Robb, T. D. (1991). *Astrophys. J.*, **372**, 329.
Cargill, P. J. and Priest, E. R. (1980). *Solar Phys.*, **65**, 251.
Cargill, P. J. and Priest, E. R. (1982). *Geophys. Astrophys. Fluid Dynamics*, **20**, 227.
Chapman, R. D., Jordan, S. D., Neupert, W. M. and Thomas, R. J. (1972). *Astrophys. J. Lett.*, **174**, L97.
Cheng, C. -C., Doschek, G. A. and Feldman, U. (1979). *Astrophys. J.*, **227**, 1037.
Chipman, E. G. (1978). *Astrophys J.*, **224**, 671.
Chiuderi, C., Einaudi, G. and Torricelli-Ciamponi, G. (1981). *Astron. Astrophys.*, **97**, 27.
Cook, J. W. and Brueckner, G. E. (1991). In *Solar Interior and Atmosphere* (Cox, A. N., Livingston, W. C. and Mathews, M. S., eds.), p. 996, The University of Arizona Press, Tucson.
Cook, J. W., Brueckner, G. E. and Bartoe, J. -D. F. (1983). *Astrophys. J. Lett.*, **270**, L89.
Cook, J. W., Brueckner, G. E., Bartoe, J. -D. F. and Socker, D. G. (1984). *Adv. Space Res.*, **4**, No. 8, 59.
Cook, J. W., Cheng, C. -C., Jacobs, V. L. and Antiochos, S. K. (1989). *Astrophys. J.*, **338**, 1176.
Cook, J. W. and Ewing, J. A. (1990). *Astrophys. J.*, **355**, 719.
Cook, J. W., Lund, P. A., Bartoe, J. -D. F., Brueckner, G. E., Dere, K. P. and Socker, D. G. (1988). In *Cool Stars, Stellar Systems, and the Sun, Lecture Notes in Physics*, 291 (Linsky, J. L. and Stencel, R. E., eds.), p. 150, Springer-Verlag, Berlin.
Cook, J. W. and Nicolas, K. R. (1979). *Astrophys. J.*, **229**, 1163.
Cox, D. P. and Tucker, W. H. (1969). *Astrophys. J.*, **157**, 1157.
Craig, I. J. D. and Brown, J. C. (1976). *Astron. Astrophys.*, **49**, 239.
Craig, I. J. D. and Brown, J. C. (1986). *Inverse Problems in Astronomy*, Adam Hilger, Bristol.
Craig, I. J. D. and McClymont, A. N. (1986). *Astrophys. J.*, **307**, 367.
Craig, I. J. D., McClymont, A. N. and Underwood, J. H. (1978). *Astron. Astrophys.*, **70**, 1.

References

de Castro, E., Fernández-Figueroa, M. J. and Rego, M. (1982). *Astron. Astrophys.*, **113**, 94.

Delache, P. (1967). *Ann. Astrophys.*, **30**, 827.

Dere, K. P. (1989). *Astrophys. J.*, **340**, 599.

Dere, K. P., Bartoe, J.-D. F. and Brueckner, G. E. (1982). *Astrophys. J.*, **259**, 366.

Dere, K. P., Bartoe, J.-D. F. and Brueckner, G. E. (1984). *Astrophys. J.*, **281**, 870.

Dere, K. P., Bartoe, J.-D. F. and Brueckner, G. E. (1986). *Astrophys. J.*, **305**, 947.

Dere, K. P., Bartoe, J.-D. F. and Brueckner, G. E. (1989a). *Solar Phys.*, **123**, 41.

Dere, K. P., Bartoe, J.-D. F., Brueckner, G. E., Cook, J. W. and Socker, D. G. (1987). *Solar Phys.*, **114**, 223.

Dere, K. P., Bartoe, J.-D. F., Brueckner, G. E., Dykton, M. D. and Van Hoosier, M. E. (1981). *Astrophys. J.*, **249**, 333.

Dere, K. P., Bartoe, J.-D. F., Brueckner, G. E. and Recely, F. (1989b). *Astrophys. J. Lett.*, **345**, L95.

Dere, K. P. and Mason, H. E. (1981). In *Solar Active Regions* (Orrall, F. Q., ed.), p. 129, Colorado Associated University Press, Boulder.

Dere, K. P., Mason, H. E., Widing, K. G. and Bhatia, A. K. (1979). *Astrophys. J. Suppl.*, **40**, 341.

Deubner, F.-L. (1975). *Astron. Astrophys.*, **44**, 371.

Deubner, F.-L. (1981). In *The Sun as a Star* (Jordan, S., ed.), p. 65, NASA SP-450.

Deubner, F.-L. and Gough, D. (1984). *Ann. Rev. Astron. Astophys.*, **22**, 593.

Doschek, G. A. (1984). *Astrophys. J.*, **279**, 446.

Doschek, G. A. (1985). In *Autoionization* (Temkin, A., ed.), p. 171, Plenum, New York.

Doschek, G. A. and Feldman, U. (1982). *Astrophys. J.*, **254**, 371.

Doschek, G. A. and Feldman, U. (1987). *Astrophys. J. Lett.*, **315**, L67.

Doschek, G. A., Feldman, U., Bhatia, A. K. and Mason, H. E. (1978a). *Astrophys. J.*, **226**, 1129.

Doschek, G. A., Feldman, U. and Bohlin, J. D. (1976a). *Astrophys. J. Lett.*, **205**, L177.

Doschek, G. A., Feldman, U. and Cohen, L. (1977). *Astrophys. J. Suppl.*, **33**, 101.

Doschek, G. A., Feldman, U., Mariska, J. T. and Linsky, J. L. (1978b). *Astrophys J. Lett.*, **226**, L35.

Doschek, G. A., Feldman, U., Van Hoosier, M. E. and Bartoe, J.-D. F. (1976b). *Astrophys. J. Suppl.*, **31**, 417.

Dowdy, J. F., Jr. (1990). *Bull. Am. Astron. Soc.*, **22**, 815.

Dowdy, J. F., Jr., Emslie, A. G. and Moore, R. L. (1987). *Solar Phys.*, **112**, 255.

Dowdy, J. F., Jr., Rabin, D. and Moore, R. L. (1986). *Solar Phys.*, **105**, 35.

Doyle, J. G., Dufton, P. L., Keenan, F. P. and Kingston, A. E. (1983). *Solar Phys.*, **89**, 243.

Doyle, J. G. and McWhirter, R. W. P. (1980). *Mon. Not. Roy. Astr. Soc.*, **193**, 947.

Dufton, P. L., Berrington, K. A., Burke, P. G. and Kingston, A. E. (1978). *Astron. Astrophys.*, **62**, 111.

Dufton, P. L., Doyle, J. G. and Kingston, A. E. (1979). *Astron. Astrophys.*, **78**, 318.

Dufton, P. L., Hibbert, A., Kingston, A. E. and Doschek, G. A. (1982). *Astrophys. J.*, **257**, 338.

Dufton, P. L., Hibbert, A., Kingston, A. E. and Doschek, G. A. (1983). *Astrophys. J.*, **274**, 420.

Dufton, P. L. and Kingston, A. E. (1980). *J. Phys. B*, **13**, 4277.

Dufton, P. L. and Kingston, A. E. (1985). *Astrophys. J.*, **289**, 844.

Dufton, P. L. and Kingston, A. E. (1987). *J. Phys. B*, **20**, 3899.

Dufton, P. L., Kingston, A. E. and Keenan, F. P. (1984). *Astrophys. J. Lett.*, **280**, L35.

Dupree, A. K. (1972). *Astrophys. J.*, **178**, 527.

Dupree, A. K. (1975). *Astrophys. J. Lett.*, **200**, L27.

Dupree, A. K., Foukal, P. V. and Jordan, C. (1976). *Astrophys. J.*, **209**, 621.

Dupree, A. K. and Goldberg, L. (1967). *Solar Phys.*, **1**, 229.

Dupree, A. K., Moore, R. T. and Shapiro, P. R. (1979). *Astrophys. J. Lett.*, **229**, L101.

Dupree, A. K. and Reimers, D. (1987). In *Exploring the Universe with the IUE Satellite* (Kondo, Y., ed), p. 321, Reidel, Dordrecht.

Dupree, A. K. and Reeves, E. M. (1971). *Astrophys. J.*, **165**, 599

Edlén, B. (1942). *Z. Astrophys.*, **22**, 30.

Evans, R. G., Jordan, C. and Wilson, R. (1975). *Mon. Not. Roy. Astr. Soc.*, **172**, 585.

Faucher, P., Masnou-Seeuws, F. and Prudhomme, M. (1980). *Astron. Astrophys.*, **81**, 137.

Feldman, U. (1983). *Astrophys. J.*, **275**, 367.

Feldman, U. (1987). *Astrophys. J.*, **320**, 426.

Feldman, U. and Behring, W. E. (1974). *Astrophys. J. Lett.*, **189**, L45.

Feldman, U., Brown, C. M., Doschek, G. A., Moore, C. E. and Rosenberg, F. D. (1976a). *J. Opt. Soc. Am.*, **66**, 853.

Feldman, U. and Doschek, G. A. (1977a). *Astrophys. J. Lett.*, **212**, L147.

Feldman, U. and Doschek, G. A. (1977b). *J. Opt. Soc. Am.*, **67**, 726.

Feldman, U. and Doschek, G. A. (1979). *Astron. Astrophys.*, **79**, 357.

Feldman, U., Doschek, G. A. and Behring, W. E. (1978a). *Space Sci. Rev.*, **22**, 191.

Feldman, U., Doschek, G. A. and Mariska, J. T. (1979). *Astrophys. J.*, **229**, 369.

Feldman, U., Doschek, G. A., Mariska, J. T., Bhatia, A. K. and Mason, H. E. (1978b). *Astrophys. J.*, **226**, 674.

Feldman, U., Doschek, G. A. and Patterson, N. P. (1976b). *Astrophys. J.*, **209**, 270.

Feldman, U., Doschek, G. A. and Rosenberg, F. D. (1977). *Astrophys. J.*, **215**, 652.

Feldman, U., Doschek, G. A. and Widing, K. G. (1978c). *Astrophys. J.*, **219**, 304.

Feldman, U., Purcell, J. D. and Dohne, B. (1987). *An Atlas of Extreme Ultraviolet Spectroheliograms from 170 to 625 Angstroms. Vol. I: Limb Features*, E. O. Hulburt Center for Space Research, Naval Research Laboratory, Washington, DC.

Feldman, U. and Widing, K. G. (1990). *Astrophys. J.*, **363**, 292.

Feldman, U., Widing, K. G. and Lund, P. A. (1990). *Astrophys. J. Lett.*, **364**, 621.

Fernández-Figueroa, M. J., de Castro, E. and Rego, M. (1981). *Astron. Astrophys.*, **99**, 141.

Fiedler, R. A. S. and Cally, P. S. (1990). *Solar Phys.*, **126**, 69.

Flower, D. R. and Nussbaumer, H. (1975a). *Astron. Astrophys.*, **42**, 265.

Flower, D. R. and Nussbaumer, H. (1975b). *Astron. Astrophys.*, **45**, 145.

Fontenla, J. M., Avrett, E. H. and Loeser, R. (1990). *Astrophys. J.*, **355**, 700.

References

Fontenla, J. M., Avrett, E. H. and Loeser, R. (1991). *Astrophys. J.*, **377**, 712.
Francis, M. H. (1981). *Solar Phys.*, **69**, 239.
Frazier, E. N. and Stenflo, J. O. (1972). *Solar Phys.*, **27**, 330.
Gabriel, A. H. (1976). *Phil. Trans. Roy. Soc. Lond. A*, **281**, 339.
Gebbie, K. B., Hill, F., Toomre, J., November, L. J., Simon, G. W., Gurman, J. B., Shine, R. A., Woodgate, B. E., Athay, R. G., Bruner, E. C., Jr., Rehse, R. A. and Tandberg-Hanssen, E. A. (1981). *Astrophys. J. Lett.*, **251**, L115.
Giacconi, R. *et al.* 1979, *Astrophys. J.*, **230**, 540.
Gilman, P. A. (1983). *Astrophys. J. Suppl.*, **53**, 243.
Giovanelli, R. G. (1949). *Mon. Not. Roy. Astr. Soc.*, **109**, 372.
Giovanelli, R. G. (1980). *Solar Phys.*, **68**, 49.
Giovanelli, R. G. (1982). *Solar Phys.*, **77**, 27.
Glass, R. and Hibbert, A. (1978). *J. Phys. B*, **11**, 2413.
Glatzmaier, G. A. (1985). *Geophys. Astrophys. Fluid Dyn.*, **31**, 137.
Golub, L., Krieger, A. S., Harvey, J. W. and Vaiana, G. S. (1977). *Solar Phys.*, **53**, 111.
Grossmann-Doerth, U. and von Uexküll, M. (1973). *Solar Phys.*, **28**, 319.
Grotrian, W. (1939). *Naturwissenschaften*, **27**, 14.
Gurevich, A. V. and Istomin, Y. N. (1979). *J. Exp. Theoret. Phys.*, **50**, 470.
Habbal, S. R., Dowdy, J. F., Jr. and Withbroe, G. L. (1990). *Astrophys. J.*, **352**, 333.
Habbal, S. R. and Withbroe, G. L. (1981). *Solar Phys.*, **69**, 77.
Harvey, K. L., Harvey, J. W. and Martin, S. F. (1975). *Solar Phys.*, **40**, 87.
Hassler, D. M., Rottman, G. J. and Orrall, F. Q. (1991). *Astrophys. J.*, **372**, 710.
Hassler, D. M., Rottman, G. J., Shoub, E. C. and Holzer, T. E. (1990). *Astrophys. J. Lett.*, **348**, L77.
Hibbert, A. (1974). *J. Phys. B*, **7**, 1417.
Hibbert, A. (1980). *J. Phys. B*, **13**, 1721.
Hollweg, J. V. (1982). *Astrophys. J.*, **257**, 345.
Hollweg, J. V. (1984). *Astrophys. J.*, **277**, 392.
Hollweg, J. V. (1985). In *Chromospheric Diagnostics and Modelling* (Lites, B. W., ed.), p. 235, National Solar Observatory, Sunspot.
Hood, A. W. and Priest, E. R. (1979). *Astron. Astrophys.*, **77**, 233.
Howard, R. and Stenflo, J. O. (1972). *Solar Phys.*, **22**, 402.
Jacobs, V. L., Davis, J., Kepple, P. C. and Blaha, M. (1977a). *Astrophys. J.*, **211**, 605.
Jacobs, V. L., Davis, J., Kepple, P. C. and Blaha, M. (1977b). *Astrophys. J.*, **215**, 690.
Jacobs, V. L., Davis, J., Rogerson, J. E. and Blaha, M. (1979). *Astrophys. J.*, **230**, 627.
Jamar, C., Macon-Hercot, P. and Praderie, F. (1976). *Astron. Astrophys.*, **52**, 373.
Jensen, E. and Orrall, F. Q. (1963). *Astrophys. J.*, **138**, 252.
Jordan, C. (1969). *Mon. Not. Roy. Astr. Soc.*, **142**, 501.
Jordan, C. (1970). *Mon. Not. Roy. Astr. Soc.*, **148**, 17.
Jordan, C. (1971). In *Highlights of Astronomy* (de Jager, C., ed.), p. 519, Reidel, Dordrecht.
Jordan, C. (1975). In *Solar Gamma-, X-, and EUV Radiation* (Kane, S. R., ed.), p. 10, Reidel, Dordrecht.
Jordan, C. (1976). *Phil. Trans. Roy. Soc. Lond. A*, **281**, 391.
Jordan, C. (1980). *Astron. Astrophys.*, **86**, 355.

Jordan, C., Ayres, T. R., Brown, A., Linsky, J. L. and Simon, T. (1987). *Mon. Not. Roy. Astr. Soc.*, **225**, 903.

Jordan, C., Brown, A., Walter, F. M. and Linsky, J. L. (1986). *Mon. Not. Roy. Astr. Soc.*, **218**, 465.

Jordan, C. and Judge, P. (1984). *Physica Scripta*, **T8**, 43.

Jordan, C. and Linsky, J. L. (1987). In *Exploring the Universe with the IUE Satellite* (Kondo, Y., ed.), p. 259, Reidel, Dordrecht.

Jordan, C. and Wilson, R. (1971). In *Physics of the Solar Corona* (Macris, C. J., ed.), p. 219, Reidel, Dordrecht.

Joselyn, J. A., Munro, R. H. and Holzer, T. E. (1979). *Astrophys. J. Suppl.*, **40**, 793.

Kanno, M. (1978). *Publ. Astron. Soc. Japan*, **30**, 581.

Kanno, M. (1979). *Publ. Astron. Soc. Japan*, **31**, 115.

Kanno, M. (1983). *Solar Phys.*, **89**, 253.

Kanno, M. and Suematsu, Y. (1982). *Publ. Astron. Soc. Japan*, **34**, 449.

Kanno, M., Suematsu, Y. and Nishikawa, T. (1984). *Solar Phys.*, **91**, 71.

Keenan, F. P. (1984). *Solar Phys.*, **91**, 27.

Keenan, F. P. (1990). *Solar Phys.*, **126**, 311.

Keenan, F. P. and Aggarwal, K. M. (1989). *Astrophys. J.*, **344**, 522.

Keenan, F. P. and Berrington, K. A. (1985). *Solar Phys.*, **99**, 25.

Keenan, F. P., Berrington, K. A., Burke, P. G., Dufton, P. L. and Kingston, A. E. (1986a). *Physica Scripta*, **34**, 216.

Keenan, F. P., Berrington, K. A., Burke, P. G., Kingston, A. E. and Dufton, P. L. (1984). *Mon. Not. Roy. Astr. Soc.*, **207**, 459.

Keenan, F. P., Cook, J. W., Dufton, P. L. and Kingston, A. E. (1989). *Astrophys. J.*, **340**, 1135.

Keenan, F. P., Dufton, P. L., Aggarwal, K. M. and Kingston, A. E. (1988). *Astrophys. J.*, **324**, 1068.

Keenan, F. P., Dufton, P. L. and Kingston, A. E. (1986b). *Astron. Astrophys.*, **169**, 319.

Keenan, F. P., Dufton, P. L. and Kingston, A. E. (1987). *Mon. Not. Roy. Astr. Soc.*, **225**, 859.

Keenan, F. P., Dufton, P. L. and Kingston, A. E. (1990). *Astrophys. J.*, **353**, 636.

Kjeldseth Moe, O. and Nicolas, K. R. (1977). *Astrophys. J.*, **211**, 579.

Klimchuk, J. A. (1986). Ph. D. Thesis, University of Colorado.

Klimchuk, J. A. (1987). *Astrophys. J.*, **323**, 368.

Klimchuk, J. A. and Mariska, J. T. (1988). *Astrophys. J.*, **328**, 334.

Kopp, R. A. (1972). *Solar Phys.*, **27**, 373.

Kozlovsky, B. -Z. and Zirin, H. (1968). *Solar Phys.*, **5**, 50.

Kraft, R. P. (1967). *Astrophys. J.*, **150**, 551.

Krieger, A. S., Vaiana, G. S. and Van Speybroeck, L. P. (1971). In *IAU Symposium 43, Solar Magnetic Fields* (R. Howard, ed.), p. 397, Reidel, Dordrecht.

Landini, M. and Monsignori Fossi, B. C. (1972). *Astron. Astrophys. Suppl. Ser.*, **7**, 291.

Landini, M. and Monsignori Fossi, B. C. (1990). *Astron. Astrophys. Suppl. Ser.*, **82**, 229.

Leibacher, J. W. and Stein, R. F. (1981). In *The Sun as a Star* (Jordan, S., ed.), p. 263, NASA SP-450.

Leighton, R. B., Noyes, R. W. and Simon, G. W. (1962). *Astrophys. J.*, **135**, 474.

References

Linsky, J. L. (1980). *Ann. Rev. Astron. Astrophys.*, **18**, 439.
Lites, B. W., Bruner, E. C., Jr., Chipman, E. G., Shine, R. A., Rottman, G. J., White, O. R. and Athay, R. G. (1976). *Astrophys. J. Lett.*, **210**, L111.
Lites, B. W. and Chipman, E. G. (1979). *Astrophys. J.*, **231**, 570.
Lites, B. W., Chipman, E. G. and White, O. R. (1982). *Astrophys. J.*, **253**, 367.
Liu, S. Y., Sheeley, N. R., Jr. and Smith, E. v. P. (1972). *Solar Phys.*, **23**, 289.
Malinovsky, M. (1975). *Astron. Astrophys.*, **43**, 101.
Mariska, J. T. (1980). *Astrophys. J.*, **235**, 268.
Mariska, J. T. (1987). *Astrophys. J.*, **319**, 465.
Mariska, J. T. (1988). *Astrophys. J.*, **334**, 489.
Mariska, J. T. and Boris, J. P. (1983). *Astrophys. J.*, **267**, 409.
Mariska, J. T., Boris, J. P., Oran, E. S., Young, T. R., Jr. and Doschek, G. A. (1982). *Astrophys. J.*, **255**, 783.
Mariska, J. T. and Dowdy, J. F. (1991). *Bull. Am. Astron. Soc.*, **23**, 1060.
Mariska, J. T., Feldman, U. and Doschek, G. A. (1978). *Astrophys. J.*, **226**, 698.
Mariska, J. T., Feldman, U. and Doschek, G. A. (1979). *Astron. Astrophys.*, **73**, 361.
Mariska, J. T. and Hollweg, J. V. (1985). *Astrophys. J.*, **296**, 746.
Mariska, J. T. and Withbroe, G. L. (1975). *Solar Phys.*, **44**, 55.
Martens, P. C. H. and Kuin, N. P. M. (1982). *Astron. Astrophys.*, **112**, 366.
Mason, H. E. and Bhatia, A. K. (1978). *Mon. Not. Roy. Astr. Soc.*, **184**, 423.
Matheson, T. D. and Noyes, R. W. (1990). *Bull. Am. Astron. Soc.*, **22**, 852.
McClymont, A. N. (1989). *Astrophys. J. Lett.*, **347**, L47.
McClymont, A. N. and Canfield, R. C. (1983). *Astrophys. J.*, **265**, 497.
McClymont, A. N. and Craig, I. J. D. (1986). *Nature*, **324**, 128.
McClymont, A. N. and Craig, I. J. D. (1987). *Astrophys J.*, **312**, 402.
McWhirter, R. W. P., Thonemann, P. C. and Wilson, R. (1975). *Astron. Astrophys.*, **40**, 63.
Meier, R. R., Widing, K. G. and Feldman, U. (1991). *Astrophys. J.*, **369**, 570.
Mein, N. (1977). *Solar Phys.*, **52**, 283.
Meyer, A. and Nussbaumer, H. (1979). *Astron. Astrophys.*, **78**, 33.
Meyer, J. -P. (1985a). *Astrophys. J. Suppl.*, **57**, 151.
Meyer, J. -P. (1985b). *Astrophys. J. Suppl.*, **57**, 173.
Mihalas, D. (1978). *Stellar Atmospheres*, 2nd edn., Freeman, San Francisco.
Moore, R. L. and Fung, P. C. W. (1972). *Solar Phys.*, **23**, 78.
Moore, R. L., Tang, F., Bohlin, J. D. and Golub, L. (1977). *Astrophys. J.*, **218**, 286.
Mühlethaler, H. P. and Nussbaumer, H. (1976). *Astron. Astrophys.*, **48**, 109.
Munro, R. H., Dupree, A. K. and Withbroe, G. L. (1971). *Solar Phys.*, **19**, 347.
Nakada, M. P. (1969). *Solar Phys.*, **7**, 302.
Nicolas, K. R., Bartoe, J. -D. F., Brueckner, G. E. and Van Hoosier, M. E. (1979). *Astrophys. J.*, **233**, 741.
Nicolas, K. R., Brueckner, G. E., Tousey, R., Tripp, D. A., White, O. R. and Athay, R. G. (1977). *Solar Phys.*, **55**, 305.
Nishikawa, T. (1983). *Solar Phys.*, **85**, 65.
Noci, G. (1981). *Solar Phys.*, **69**, 63.

Noci, G., Spadaro, D., Zappalà, R. A. and Zuccarello, F. (1988). *Astron. Astrophys.*, **198**, 311.

Noyes, R. W., Hartmann, L. W., Baliunas, S. L., Duncan, D. K. and Vaughn, A. H. (1984). *Astrophys. J.*, **279**, 763.

Nussbaumer, H. and Storey, P. J. (1975). *Astron. Astrophys.*, **44**, 321.

Nussbaumer, H. and Storey, P. J. (1979a). *Astron. Astrophys.*, **71**, L5.

Nussbaumer, H. and Storey, P. J. (1979b). *Astron. Astrophys.*, **74**, 244.

Nussbaumer, H. and Storey, P. J. (1982). *Astron. Astrophys.*, **115**, 205.

Osterbrock, D. E. (1961). *Astrophys. J.*, **134**, 347.

Osterbrock, D. E. (1974). *Astrophysics of Gaseous Nebulae*, p. 110, W. H. Freeman, San Francisco.

Owocki, S. P. and Scudder, J. D. (1983). *Astrophys. J.*, **270**, 758.

Pallavicini, R., Peres, G., Serio, S., Vaiana, G. S., Golub, L. and Rosner, R. (1981). *Astrophys. J.*, **247**, 692.

Parker, E. N. (1988). *Astrophys. J.*, **330**, 474.

Parker, E. N. (1989). *Solar Phys.*, **121**, 271.

Peres, G., Rosner, R., Serio, S. and Vaiana, G. S. (1982). *Astrophys. J.*, **252**, 791.

Pneuman, G. W. and Kopp, R. A. (1977). *Astron. Astrophys.*, **55**, 305.

Pneuman, G. W. and Kopp, R. A. (1978). *Solar Phys.*, **57**, 49.

Porter, J. G. and Dere, K. P. (1991). *Astrophys. J.*, **370**, 775.

Porter, J. G., Moore, R. L., Reichmann, E. J., Engvold, O. and Harvey, K. L. (1987). *Astrophys. J.*, **323**, 380.

Pottasch, S. R. (1963). *Astrophys. J.*, **137**, 945.

Pottasch, S. R. (1964). *Space Sci. Rev.*, **3**, 816.

Priest, E. R. (1982). *Solar Magnetohydrodynamics*, Reidel, Dordrecht.

Rabin, D. (1986). In *Coronal and Prominence Plasmas* (Poland, A., ed.), p. 135, NASA CP 2442.

Rabin, D. (1991). *Astrophys. J.*, **383**, 407.

Rabin, D. and Moore, R. L. (1984). *Astrophys. J.*, **285**, 359.

Raymond, J. C. (1979). Personal communication.

Raymond, J. C. (1990). *Astrophys. J.*, **365**, 387.

Raymond, J. C. and Doyle, J. G. (1981a). *Astrophys. J.*, **245**, 1141.

Raymond, J. C. and Doyle, J. G. (1981b). *Astrophys, J.*, **247**, 686.

Raymond, J. C. and Dupree, A. K. (1978). *Astrophys. J.*, **222**, 379.

Raymond, J. C. and Foukal, P. (1982). *Astrophys. J.*, **253**, 323.

Raymond, J. C. and Smith, B. W. (1977). *Astrophys. J. Suppl.*, **35**, 419.

Reeves, E. M. (1976). *Solar Phys.*, **46**, 53.

Reeves, E. M., Foukal, P. V., Huber, M. C. E., Noyes, R. W., Schmahl, E. J., Timothy, J. G., Vernazza, J. E. and Withbroe, G. L. (1974). *Astrophys. J. Lett.*, **188**, L27.

Reeves, E. M., Vernazza, J. E. and Withbroe, G. L. (1976). *Phil. Trans. Roy. Soc. Lond. A*, **281**, 319.

Rosner, R., Tucker, W. H. and Vaiana, G. S. (1978). *Astrophys. J.*, **220**, 643.

Ross, J. E. and Aller, L. H. (1976). *Science*, **191**, 1223.

Rottman, G. J., Hassler, D. M., Jones, M. D. and Orrall, F. Q. (1990). *Astrophys. J.*, **358**, 693.

References

Roussel-Dupré, D. and Shine, R. A. (1982). *Solar Phys.*, **77**, 329.
Roussel-Dupré, R. (1980a). *Astrophys. J.*, **241**, 402.
Roussel-Dupré, R. (1980b). *Solar Phys.*, **68**, 243.
Roussel-Dupré, R. (1980c). *Solar Phys.*, **68**, 265.
Roussel-Dupré, R. (1981). *Astrophys J.*, **243**, 329.
Roussel-Dupré, R. and Beerman, C. (1981). *Astrophys. J.*, **250**, 408.
Roussel-Dupré, R., Francis, M. H. and Billings, D. E. (1979). *Mon. Not. Roy. Astr. Soc.*, **187**, 9.
Sandlin, G. D., Bartoe, J. -D. F., Brueckner, G. E., Tousey, R. and Van Hoosier, M. E. (1986). *Astrophys. J. Suppl.*, **61**, 801.
Schatzman, E. (1949). *Ann. d'Astrophys.*, **12**, 203.
Schmahl, E. J. and Orrall, F. Q. (1979). *Astrophys. J. Lett.*, **231**, L41.
Schmieder, B. (1978). *Solar Phys.*, **57**, 245.
Schmieder, B. and Mein, N. (1980). *Astron. Astrophys.*, **84**, 99.
Schrijver, C. J., Zwann, C., Maxson, C. W. and Noyes, R. W. (1985). *Astron. Astrophys.*, **149**, 123.
Schwarzschild, M. (1948). *Astrophys. J.*, **107**, 1.
Schwenn, R. (1983). In *Solar Wind Five* (Neugebauer, M., ed.), p. 489, NASA CP 2280.
Seaton, M. J. (1964). *Mon. Not. Roy. Astr. Soc.*, **127**, 191.
Serio, S., Peres, G., Vaiana, G. S., Golub, L. and Rosner, R. (1981). *Astrophys. J.*, **243**, 288.
Sheeley, N. R., Jr. and Golub, L. (1979). *Solar Phys.*, **63**, 119.
Shine, R., Gerola, H. and Linsky, J. L. (1975). *Astrophys. J. Lett.*, **202**, L1.
Shine, R. A., Roussel-Dupré, D., Bruner, E. C., Jr., Chipman, E. G., Lites, B. W., Rottman, G. J., Athay, R. G. and White, O. R. (1976). *Astrophys. J. Lett.*, **210**, L107.
Shoub, E. C. (1983). *Astrophys. J.*, **266**, 339.
Shull, J. M. and Van Steenberg, M. (1982a). *Astrophys. J. Suppl.*, **48**, 95.
Shull, J. M. and Van Steenberg, M. (1982b). *Astrophys. J. Suppl.*, **49**, 351.
Siarkowski, M. (1983). *Solar Phys.*, **84**, 131.
Simon, T., Herbig, G. and Boesgaard, A. M. (1985). *Astrophys. J.*, **293**, 551.
Skumanich, A., Smythe, C. and Frazier, E. N. (1975). *Astrophys. J.*, **200**, 747.
Soderblom, D. R. (1983). *Astrophys. J. Suppl.*, **53**, 1.
Spitzer, L. (1962). *Physics of Fully Ionized Gases*, 2nd edn, Interscience, New York.
Stein, R. F. and Leibacher, J. (1974). *Ann. Rev. Astron. Astrophys.*, **12**, 407.
Stein, R. F. and Leibacher, J. (1981). In *The Sun as a Star* (Jordan, S., ed.), p. 289, NASA SP-450.
Stencel, R. E. and Mullan, D. J. (1980). *Astrophys. J.*, **238**, 221.
Stenflo, J. O. (1973). *Solar Phys.*, **32**, 41.
Stenflo, J. O. (1989). *Astron. Astrophys. Rev.*, **1**, 3.
Sterling, A. C. and Hollweg, J. V. (1988). *Astrophys. J.*, **327**, 950.
Sterling, A. C. and Mariska, J. T. (1990). *Astrophys. J.*, **349**, 647.
Sterling, A. C., Mariska, J. T., Shibata, K. and Suematsu, Y. (1991). *Astrophys. J.*, **381**, 313.

Sturrock, P. A., Dixon, W. W., Klimchuk, J. A. and Antiochos, S. K. (1990). *Astrophys. J. Lett.*, **356**, L31.
Suematsu, Y., Shibata, K., Nishikawa, T. and Kitai, R. (1982). *Solar Phys.*, **75**, 99.
Summers, H. P. (1972). *Mon. Not. Roy. Astr. Soc.*, **158**, 255.
Summers, H. P. (1974). *Mon. Not. Roy. Astr. Soc.*, **169**, 663.
Sylwester, J., Schrijver, J. and Mewe, R. (1980). *Solar Phys.*, **67**, 285.
Tandberg-Hanssen, E. and Emslie, A. G. (1988). *The Physics of Solar Flares*, Cambridge University Press, Cambridge.
Torricelli-Ciamponi, G., Einaudi, G. and Chiuderi, C. (1982). *Astron. Astrophys.*, **105**, L1.
Trubnikov, B. A. (1965). In *Reviews of Plasma Physics*, Vol I (Leontovich, M. A., ed.), p. 105, Consultants Bureau, New York.
Tucker, W. H. and Koren, M. (1971). *Astrophys. J.*, **168**, 283.
Tworkowski, A. S. (1975). *Astrophys. Lett.*, **17**, 27.
Tworkowski, A. S. (1980). *Mon. Not. Roy. Astr. Soc.*, **190**, 287.
Ulmschneider, P. (1970). *Astron. Astrophys.*, **4**, 144.
Ulrich, R. K. (1970). *Astrophys. J.*, **162**, 993.
Van Regemorter, H. (1962). *Astrophys. J.*, **136**, 906.
Vernazza, J. E., Avrett, E. H. and Loeser, R. (1981). *Astrophys. J. Suppl.*, **45**, 635.
Vernazza, J. E., Foukal, P. V., Huber, M. C. E., Noyes, R. W., Reeves, E. M., Schmahl, E. J., Timothy, J. G. and Withbroe, G. L. (1975). *Astrophys. J. Lett.*, **199**, L123.
Vernazza, J. E. and Mason, H. E. (1978). *Astrophys. J.*, **226**, 720.
Vernazza, J. E. and Reeves, E. M. (1978). *Astrophys. J. Suppl.*, **37**, 485.
Vesecky, J. F., Antiochos, S. K. and Underwood, J. H. (1979). *Astrophys. J.*, **233**, 987.
Vitz, R. C., Weiser, H., Moos, H. W., Weinstein, A. and Worden, E. S. (1976). *Astrophys. J. Lett.*, **205**, L35.
Walker, A. B. C., Jr., Barbee, T. W., Jr., Hoover, R. B. and Lindblom, J. F. (1988). *Science*, **241**, 1781.
Wallenhorst, S. G. (1982). *Solar Phys.*, **77**, 167.
Walter, F. M. (1981). *Astrophys. J.*, **245**, 677.
Walter, F. M., Linsky, J. L., Bowyer, C. S. and Garmire, G. (1980). *Astrophys. J. Lett.*, **236**, L137.
White, O. R. and Athay, R. G. (1979). *Astrophys. J. Suppl.*, **39**, 347.
Widing, K. G., Doyle, J. G., Dufton, P. L. and Kingston, A. E. (1982). *Astrophys. J.*, **257**, 913.
Widing, K. G. and Feldman, U. (1989). *Astrophys. J.*, **344**, 1046.
Wilson, O. C. (1963). *Astrophys. J.*, **138**, 832.
Withbroe, G. L. (1970a). *Solar Phys.*, **11**, 42.
Withbroe, G. L. (1970b). *Solar Phys.*, **11**, 208.
Withbroe, G. L. (1971). *Solar Phys.*, **18**, 458.
Withbroe, G. L. (1975). *Solar Phys.*, **45**, 301.
Withbroe, G. L. (1983). *Astrophys. J.*, **267**, 825.
Withbroe, G. L. (1988). *Astrophys. J.*, **325**, 442.
Withbroe, G. L. and Mariska, J. T. (1976). *Solar Phys.*, **48**, 21.

References

Woods, D. T., Holzer, T. E. and MacGregor, K. B. (1990). *Astrophys. J.*, **355**, 295.
Yeh, T. (1977). *Solar Phys.*, **55**, 241.

AUTHOR INDEX

Acton, L. W., 65, 66, 68, 257
Aggarwal, K. M., 99, 105, 112, 256, 262
Allen, C. W., 106, 256
Allen, J. W., 29, 256
Aller, L. H., 34, 36, 84, 256, 264
Almleaky, Y. M., 108, 256
Anders, E., 34, 35, 256
Ando, H., 153, 256
Antiochos, S. K., 37, 39, 205, 214, 219, 220, 227, 228, 256, 258, 266
Arnaud, M., 29, 40, 112, 256
Artzner, G., 118, 257
Athay, R. G., 2, 3, 14, 15, 38, 41, 60, 62, 64, 69, 78, 111, 119, 120, 121, 122, 124, 132, 133, 134, 135, 136, 137, 154, 155, 167, 171, 176, 213, 219, 223, 230, 237, 256, 261, 263, 265, 266
Avrett, E. H., 11, 38, 60, 67, 68, 106, 172, 202, 232, 233, 234, 235, 256, 257, 260, 261, 266
Ayres, T. R., 239, 241, 242, 243, 244, 245, 246, 257, 262

Baliunas, S. L., 247, 264
Barbee, T. W., Jr., 71, 266
Bartoe, J. -D. F., 6, 7, 51, 52, 54, 55, 58, 61, 63, 64, 66, 71, 83, 89, 91, 92, 96, 97, 111, 116, 117, 118, 119, 120, 121, 122, 129, 130, 132, 134, 135, 136, 137, 138, 140, 141, 143, 144, 145, 147, 148, 149, 150, 156, 158, 160, 176, 177, 178, 180, 194, 204, 257, 258, 259, 263, 265
Baum, W. A., 4, 257
Beckers, J. M., 62, 63, 138, 150, 257
Beerman, C., 184, 265
Behring, W. E., 112, 118, 260
Bely, O., 24, 257
Berrington, K. A., 23, 86, 87, 88, 89, 257, 259, 262
Bhadra, K., 97, 257
Bhatia, A. K., 97, 99, 103, 104, 105, 257, 259, 260, 263
Biermann, L., 3, 153, 257
Billings, D. E., 121, 265

Blaha, M., 23, 29, 257, 261
Boesgaard, A. M., 246, 247, 265
Bohlin, J. D., 12, 130, 131, 132, 140, 151, 257, 259, 263
Boland, B. C., 116, 118, 119, 124, 126, 257
Bonnet, R. M., 65, 66, 68, 118, 257
Book, D. L., 115, 257
Boris, J. P., 161, 214, 215, 220, 257, 263
Borrini, G., 161, 257
Bowen, I. S., 2, 257
Bowyer, C. S., 243, 266
Boyd, T. J. M., 199, 257
Braginskii, S. I., 182, 201, 257
Bray, R. J., 62, 257
Breneman, H. H., 36, 257
Brown, A., 241, 242, 243, 244, 245, 246, 257, 262
Brown, C. M., 83, 260
Brown, J. C., 77, 78, 108, 256, 258
Brown, W. A., 65, 66, 68, 257
Brueckner, G. E., 6, 7, 51, 52, 54, 55, 61, 63, 64, 66, 71, 83, 89, 91, 92, 97, 111, 116, 119, 120, 121, 122, 129, 130, 132, 134, 135, 136, 137, 138, 140, 141, 143, 144, 145, 146, 147, 148, 149, 150, 156, 158, 160, 176, 177, 178, 179, 180, 204, 257, 258, 259, 263, 265
Bruner, E. C., Jr., 65, 66, 68, 119, 121, 123, 124, 127, 128, 133, 134, 135, 137, 152, 156, 257, 258, 261, 263, 265
Bruner, M. E., 76, 258
Burke, P. G., 23, 86, 87, 88, 89, 257, 259, 262
Burton, W. M., 63, 258
Byerley, A., 127, 258

Cally, P. S., 213, 231, 235, 236, 258, 260
Canfield, R. C., 37, 39, 224, 230, 263
Cargill, P. J., 214, 258
Chapman, R. D., 61, 258
Cheng, C. -C., 37, 39, 118, 119, 258
Chipman, E. G., 121, 133, 154, 155, 258, 263, 265
Chiuderi, C., 205, 258, 266

Author Index

Cohen, L., 122, 259
Cook, J. W., 37, 39, 54, 55, 63, 64, 86, 87, 88, 91, 103, 105, 107, 119, 120, 121, 122, 138, 143, 144, 145, 148, 156, 161, 176, 177, 178, 179, 204, 258, 259, 262
Cox, D. P., 37, 230, 258
Craig, I. J. D., 77, 78, 204, 214, 215, 218, 258

Davis, J., 29, 261
de Castro, E., 241, 259, 260
Decaudin, M., 65, 66, 68, 257
Delache, P., 82, 259
Dere, K. P., 54, 55, 61, 63, 64, 66, 78, 84, 97, 99, 119, 120, 121, 122, 129, 130, 132, 134, 135, 136, 137, 138, 140, 141, 143, 144, 145, 146, 147, 148, 150, 156, 158, 160, 176, 177, 178, 180, 204, 256, 258, 259, 264
Deubner, F. -L., 153, 259
Dixon, W. W., 219, 220, 266
Dohne, B., 105, 260
Doschek, G. A., 22, 41, 44, 46, 47, 48, 55, 58, 59, 60, 63, 68, 83, 84, 89, 91, 92, 93, 96, 97, 98, 99, 101, 102, 103, 104, 105, 108, 110, 111, 112, 116, 117, 118, 119, 121, 122, 124, 126, 127, 130, 131, 132, 138, 140, 160, 161, 162, 174, 175, 177, 178, 181, 194, 220, 257, 258, 259, 260, 263
Dowdy, J. F., Jr., 68, 69, 73, 130, 132, 211, 212, 218, 253, 259, 261, 263
Doyle, J. G., 76, 79, 80, 81, 86, 88, 89, 168, 169, 181, 192, 193, 209, 259, 264, 266
Dufton, P. L., 23, 86, 87, 88, 89, 91, 92, 93, 97, 98, 105, 110, 111, 160, 162, 183, 257, 259, 260, 262, 266
Duncan, D. K., 247, 264
Dupree, A. K., 29, 84, 86, 87, 111, 161, 162, 163, 167, 171, 172, 238, 240, 256, 260, 263, 264
Dyer, E. P., 116, 118, 119, 126, 257
Dykton, M. D., 61, 158, 160, 259

Edlén, B., 2, 257, 260
Einaudi, G., 205, 258, 266
Emslie, A. G., 199, 211, 212, 259, 266
Engvold, O., 147, 148, 150, 151, 244, 257, 264
Evans, R. G., 238, 260
Ewing, J. A., 64, 258

Faucher, P., 24, 257, 260
Feldman, U., 41, 46, 47, 55, 58, 59, 60, 63, 68, 83, 96, 97, 99, 102, 103, 104, 105, 110, 111, 112, 116, 117, 118, 119, 121, 122, 124, 126, 127, 130, 131, 132, 138, 140, 160, 174, 175, 177, 178, 181, 194, 195, 257, 258, 259, 260, 263, 266
Fernández-Figueroa, M. J., 241, 259, 260

Fiedler, R. A. S., 213, 235, 260
Firth, J. G., 116, 118, 119, 124, 126, 257
Flower, D. R., 97, 109, 260
Fontenla, J. M., 232, 233, 260, 261
Foukal, P. V., 52, 61, 86, 87, 173, 260, 264, 266
Francis, M. H., 121, 160, 163, 261, 265
Frazier, E. N., 54, 62, 67, 261, 265
Fung, P. C. W., 205, 263

Gabriel, A. H., 67, 68, 116, 118, 119, 124, 126, 190, 193, 195, 205, 210, 213, 257, 261
Garmire, G., 243, 266
Gebbie, K. B., 133, 134, 135, 137, 261
Gerola, H., 82, 265
Giacconi, R., 242, 261
Gilman, P. A., 247, 261
Giovanelli, R. G., 3, 67, 68, 171, 261
Glass, R., 89, 261
Glatzmaier, G. A., 247, 261
Goldberg, L., 167, 260
Golub, L., 71, 148, 151, 205, 261, 263, 264, 265
Gough, D., 153, 259
Gouttebroze, P., 118, 257
Grevesse, N., 34, 35, 256
Grossmann-Doerth, U., 64, 261
Grotrian, W., 2, 261
Gurevich, A. V., 183, 261
Gurman, J. B., 120, 133, 134, 135, 136, 137, 256, 261

Habbal, S. R., 73, 261
Hartmann, L. W., 247, 264
Harvey, J. W., 71, 261
Harvey, K. L., 71, 147, 148, 150, 151, 261, 264
Hassler, D. M., 118, 119, 123, 127, 130, 132, 218, 235, 264
Henry, R. J. W., 97, 257
Henze, W., 120, 135, 136, 137, 256
Herbig, G., 246, 247, 265
Hibbert, A., 88, 89, 91, 92, 93, 97, 98, 162, 260, 261
Hill, F., 133, 134, 135, 137, 261
Hollweg, J. V., 128, 141, 223, 261, 263, 265
Holzer, T. E., 118, 119, 123, 127, 159, 193, 219, 227, 229, 256, 261, 262, 267
Hood, A. W., 224, 261
Hoover, R. B., 71, 266
Howard, R., 67, 261
Huber, M. C. E., 52, 61, 264, 266

Istomin, Y. N., 183, 261

Jacobs, V. L., 29, 37, 39, 258, 261
Jamar, C., 238, 261
Jensen, E., 54, 244, 245, 257, 261
Johnson, F. S., 4, 257

Jones, B. B., 116, 118, 119, 124, 126, 257
Jones, M. D., 130, 132, 264
Jordan, C., 29, 63, 75, 78, 84, 86, 87, 116, 118, 119, 124, 126, 172, 190, 191, 192, 193, 229, 238, 240, 241, 242, 243, 244, 245, 246, 257, 258, 260, 261, 262
Jordan, S. D., 61, 258
Joselyn, J. A., 159, 262
Jouchoux, A., 118, 257
Judge, P., 240, 262

Kanno, M., 174, 181, 262
Keenan, F. P., 86, 87, 88, 89, 91, 105, 110, 111, 112, 160, 183, 259, 260, 262
Kepple, P. C., 29, 261
Kingston, A. E., 23, 86, 87, 88, 89, 91, 92, 93, 97, 98, 105, 110, 111, 160, 162, 183, 257, 259, 260, 262, 266
Kitai, R., 223, 266
Kjeldseth Moe, O., 58, 91, 119, 262
Klimchuk, J. A., 132, 137, 219, 220, 225, 226, 231, 262, 266
Kopp, R. A., 138, 139, 147, 196, 212, 213, 219, 262, 264
Koren, M., 37, 266
Kozlovsky, B. -Z., 31, 262
Kraft, R. P., 246, 247, 262
Krieger, A. S., 71, 148, 261, 262
Kuin, N. P. M., 227, 263
Kurokawa, H., 148, 258

Landini, M., 29, 262
Leibacher, J. W., 118, 153, 257, 262, 265
Leighton, R. B., 153, 262
Lemaire, P., 118, 257
Lindblom, J. F., 71, 266
Linsky, J. L., 82, 102, 238, 239, 240, 241, 242, 243, 244, 245, 246, 257, 259, 262, 263, 265, 266
Lites, B. W., 121, 133, 153, 154, 263, 265
Liu, S. Y., 64, 263
Loeser, R., 11, 38, 60, 67, 68, 106, 172, 202, 230, 232, 233, 234, 235, 260, 261, 266
Loughhead, R. E., 62, 257
Lund, P. A., 83, 143, 144, 145, 258, 260

MacGregor, K. B., 193, 227, 229, 267
Macon-Hercot, P., 238, 261
Malinovsky, M., 22, 263
Mariska, J. T., 40, 57, 58, 59, 60, 76, 82, 99, 102, 118, 122, 124, 130, 132, 138, 161, 174, 175, 177, 178, 180, 194, 214, 215, 216, 218, 220, 221, 222, 225, 226, 231, 257, 259, 260, 262, 263, 266
Marstad, N. C., 239, 257
Martens, P. C. H., 227, 263
Martin, S. F., 71, 261
Masnou-Seeuws, F., 24, 260

Mason, H. E., 78, 84, 93, 97, 99, 103, 104, 105, 259, 260, 263, 266
Matheson, T. D., 61, 263
Maxson, C. W., 53, 56, 265
McCabe, M., 148, 258
McClymont, A. N., 37, 39, 204, 214, 215, 218, 224, 230, 231, 232, 258, 263
McWhirter, R. W. P., 37, 76, 86, 116, 118, 119, 123, 124, 126, 127, 128, 201, 205, 230, 257, 258, 259, 263
Meier, R. R., 83, 263
Mein, N., 153, 263, 265
Mewe, R., 78, 266
Meyer, A., 82, 263
Meyer, J. -P., 34, 35, 36, 263
Mihalas, D., 116, 263
Moe, O. K., 116, 258
Monk, P., 116, 118, 119, 124, 126, 257
Monsignori Fossi, B. C., 29, 262
Moore, C. E., 83, 260
Moore, R. L., 68, 147, 148, 150, 151, 205, 211, 212, 237, 253, 259, 263, 264
Moore, R. T., 161, 163, 260
Moos, H. W., 238, 266
Mühlethaler, H. P., 86, 263
Mullan, D. J., 240, 265
Munro, R. H., 84, 159, 262, 263

Nakada, M. P., 82, 263
Neupert, W. M., 61, 258
Nicolas, K. R., 58, 63, 86, 87, 88, 89, 91, 92, 102, 103, 105, 107, 111, 119, 122, 161, 258, 262, 263
Nishikawa, T., 181, 223, 262, 263, 266
Noci, G., 83, 161, 214, 227, 228, 256, 257, 263, 264
November, L. J., 133, 134, 135, 137, 261
Noyes, R. W., 52, 53, 56, 61, 153, 247, 262, 263, 264, 265, 266
Nussbaumer, H., 26, 82, 86, 88, 89, 96, 97, 109, 260, 263, 264

Oberly, J. J., 4, 257
Oran, E. S., 161, 220, 263
Orrall, F. Q., 54, 68, 130, 132, 181, 218, 235, 261, 264, 265
Osaki, Y., 153, 256
Osterbrock, D. E., 3, 98, 153, 264
Owocki, S. P., 183, 264

Pallavicini, R., 205, 264
Parker, E. N., 219, 264
Patterson, N. P., 55, 119, 121, 126, 260
Peres, G., 205, 220, 265
Pneuman, G. W., 138, 212, 213, 219, 264
Porter, J. G., 147, 148, 150, 151, 264
Pottasch, S. R., 74, 76, 82, 167, 264
Praderie, F., 238, 261
Priest, E. R., 203, 214, 224, 258, 261, 264
Prudhomme, M., 24, 260

Author Index

Purcell, J. D., 6, 12, 105, 257, 260

Rabin, D., 68, 211, 237, 253, 259, 264
Raymond, J. C., 37, 38, 39, 76, 79, 80, 81, 161, 162, 168, 169, 173, 181, 192, 193, 209, 220, 227, 228, 264
Recely, F., 130, 259
Reeves, E. M., 7, 52, 53, 54, 55, 56, 61, 62, 68, 71, 73, 79, 80, 83, 111, 112, 211, 260, 264, 266
Rego, M., 241, 259, 260
Rehse, R. A., 133, 134, 135, 137, 261
Reichmann, E. J., 147, 148, 150, 151, 264
Reimers, D., 240, 260
Ridgeley, A., 63, 258
Robb, T. D., 231, 258
Rockwood, C. C., 4, 257
Rogerson, J. E., 29, 261
Rosenberg, F. D., 83, 99, 260
Rosner, R., 37, 204, 205, 215, 220, 264, 265
Ross, J. E., 36, 264
Rothenflug, R., 29, 40, 112, 256
Rottman, G. J., 118, 119, 121, 123, 127, 130, 132, 133, 218, 235, 261, 263, 264, 265
Roussel-Dupré, D., 121, 130, 131, 132, 136, 265
Roussel-Dupré, R., 121, 183, 184, 265

Sanderson, J. J., 199, 257
Sandlin, G. D., 7, 83, 265
Schatzman, E., 153, 265
Schmahl, E. J., 52, 61, 68, 181, 264, 265, 266
Schmieder, B., 153, 265
Schrijver, C. J., 53, 56, 265
Schrijver, J., 78, 266
Schwarzschild, M., 3, 153, 265
Schwenn, R., 149, 265
Scudder, J. D., 183, 264
Seaton, M. J., 24, 265
Scrio, S., 205, 220, 264, 265
Shapiro, P. R., 161, 163, 260
Sheeley, N. R., Jr., 12, 64, 71, 148, 257, 263, 265
Shibata, K., 223, 265, 266
Shine, R. A., 82, 120, 121, 130, 131, 132, 133, 134, 136, 137, 256, 261, 263, 265
Shoub, E. C., 118, 119, 123, 127, 183, 261, 265
Shull, J. M., 29, 30, 31, 112, 157, 265
Siarkowski, M., 78, 265
Simon, G. W., 133, 134, 135, 137, 153, 261, 262
Simon, T., 241, 242, 243, 244, 245, 246, 247, 257, 262, 265
Skumanich, A., 54, 62, 118, 257, 265
Smith, B. W., 37, 264
Smith, E. v. P., 64, 263
Smythe, C., 54, 62, 265

Socker, D. G., 54, 55, 64, 119, 120, 121, 122, 138, 143, 144, 145, 148, 156, 176, 177, 178, 204, 258, 259
Soderblom, D. R., 246, 265
Spadaro, D., 83, 264
Spitzer, L., 198, 201, 202, 236, 265
Stein, R. F., 262, 265
Stencel, R. E., 240, 245, 257, 265
Stenflo, J. O., 67, 197, 261, 265
Sterling, A. C., 223, 265
Stone, E. C., 36, 257
Storey, P. J., 26, 88, 89, 96, 264
Strain, C. V., 4, 257
Sturrock, P. A., 219, 220, 266
Suematsu, Y., 181, 223, 262, 265, 266
Summers, H. P., 29, 266
Sweet, P. A., 108, 256
Sylwester, J., 78, 266

Tandberg-Hanssen, E. A., 133, 134, 135, 137, 199, 261, 266
Tang, F., 151, 263
Thomas, R. J., 61, 258
Thomas, R. N., 3, 256
Thonemann, P. C., 37, 201, 205, 230, 263
Timothy, J. G., 52, 61, 264, 266
Toomre, J., 133, 134, 135, 137, 261
Torricelli-Ciamponi, G., 205, 258, 266
Tousey, R., 4, 6, 7, 12, 83, 111, 122, 257, 263, 265
Tripp, D. A., 111, 122, 263
Trubnikov, B. A., 115, 266
Tucker, W. H., 37, 204, 215, 230, 258, 264, 266
Turner, R. F., 116, 118, 119, 124, 126, 257
Tworkowski, A. S., 82, 184, 266

Ufford, C. W., 84, 256
Ulmschneider, P., 171, 201, 266
Ulrich, R. K., 153, 266
Underwood, J. H., 204, 205, 258, 266

Vaiana, G. S., 37, 71, 148, 204, 205, 215, 220, 261, 262, 264, 265
Van Hoosier, M. E., 7, 12, 58, 61, 83, 89, 91, 92, 96, 97, 111, 116, 117, 118, 158, 160, 194, 257, 259, 263, 265
Van Regemorter, H., 23, 26, 74, 160, 266
Van Speybroeck, L. P., 148, 262
Van Steenberg, M., 29, 30, 31, 112, 157, 265
Van Vleck, J. H., 84, 256
Vaughn, A. H., 247, 264
Vernazza, J. E., 7, 11, 38, 52, 60, 61, 67, 68, 71, 79, 80, 83, 93, 97, 106, 112, 172, 202, 230, 233, 234, 235, 264, 266
Vesecky, J. F., 205, 266
Vial, J. C., 118, 257
Vidal-Madjar, A., 118, 257
Vitz, R. C., 238, 266

Vogel, S. N., 12, 257
von Uexküll, M., 64, 261

Walker, A. B. C., Jr., 71, 266
Wallenhorst, S. G., 213, 266
Walter, F. M., 241, 243, 262, 266
Weinstein, A., 238, 266
Weiser, H., 238, 266
White, O. R., 111, 119, 121, 122, 124, 133, 154, 155, 256, 263, 265, 266
Widing, K. G., 46, 47, 83, 88, 89, 99, 259, 260, 263, 266
Wilson, O. C., 246, 266
Wilson, R., 37, 63, 75, 78, 127, 172, 201, 205, 230, 238, 258, 260, 262, 263

Withbroe, G. L., 52, 57, 60, 61, 63, 71, 73, 78, 82, 84, 149, 174, 175, 180, 261, 263, 264, 266
Woodgate, B. E., 133, 134, 135, 137, 261
Woods, D. T., 193, 227, 229, 267
Worden, E. S., 238, 266

Yeh, T., 214, 267
Young, T. R., Jr., 161, 220, 263

Zappalà, R. A., 83, 264
Zirin, H., 31, 262
Zuccarello, F., 83, 264
Zwann, C., 53, 56, 265

SUBJECT INDEX

abundances, elemental, 34–37
  absolute, 82
  and diffusion, 185
  and radiative losses, 231
  coronal, 35
  critical compilations, 34
  determinations, 81–84
  differences, 35
  gradients, 166
  photospheric, 35
  variations, 82–84
abundances, hydrogen to electron, 36
acoustic waves, *see* waves
Alfvén speed, 198
Alfvén waves, *see* waves
Ampére's law, 197
atomic processes, 17–19
  timescales, 18
autoionization, 19, 28

barometric equation, 167
beta, magnetic field, 197
Boltzmann factor, and temperature diagnostics, 109
bright points, 71–73, 148
  and explosive events, 150–151

cell emission measure, 80
chromosphere, 13
chromospheric network, 9
collision strength, 21, 22
  approximate formula, 23
collisional deexcitation, 18, 23
collisional excitation, 18
  proton, 24
  rate coefficient, 21
  rates, 21–24
collisional ionization, 18
conduction, thermal
  cross field, 201–202, 236–237, 251
conduction, turbulent, 235–236
conductive flux, thermal, 171–172
  relation to pressure, 172
continuum emission, UV, 3, 5, 6

contribution function, 25
  width, 75
convection, and surface magnetic fields, 246
cooling
  and ionization balance, 160
  and recombination, 158
corona, 14
Coulomb logarithm, 115, 198
current density, electric, 197

density
  estimates, 87, 89, 91, 94, 95, 96, 98, 99, 102, 103, 104
  quiet-Sun variations, 107
  transition region, 105–108, 248
density diagnostics, 43–50
  aluminum isoelectronic sequence, 97–98
  and departures from ionization equilibrium, 161–163
  based on volume estimates, 43
  beryllium isoelectronic sequence, 84–89
  boron isoelectronic sequence, 92–97
  carbon isoelectronic sequence, 98–99
  four-level atom, 47–49
  magnesium isoelectronic sequence, 89–92
  multicomponent plasma, 108
  nitrogen isoelectronic sequence, 99
  stellar, 241
  temperature dependence, 87
  temperature sensitivity, 49
  three-level atom, 43–47
  using different ions, 99–105
detailed balance, 23–24
dielectronic recombination, 19, 28, 33
diffusion, thermal, 184–185, 232–235, 251
  effect on abundances, 82
  velocities, 235
Doppler broadening, 114–115
Doppler shifts, 129–141, 249
  and line widths, 136
  and mass balance, 137–139
  and network, 133–135
  center to limb behavior, 136–137

explosive events, 143
observations, 129–133
stellar, 245
temporal behavior, 133
Doppler width, 115
downflows
  and recombination, 158
  explanations, 219
  heating as driver, 214–219
  in cooling loops, 221–223
  ionization balance in, 160
  models, 212–213
  stellar transition regions, 245
dynamo model, 246

electric current heating, 237
electron density, see density
electron impact ionization, 28
emission line profile, 114–116
  and Doppler broadening, 136
  explosive events, 145–146
  temperature dependence, 117–119
emission lines
  list, 8
  non-Gaussian line profiles, 141–151
  nonthermal broadening and mechanical energy, 192
  temperature of formation, 7, 8, 75
emission measure, 249
  calculating, 74–79
  cool loop, 228
  definition, 75
  departures from equilibrium, 160
  differential, 77
  elemental abundances, 81
  explosive events, 143, 146
  features, 248
  from static models, 209–212
  generalized, 77
  gradient and energy balance, 191
  in diverging geometries, 211–212
  iterative techniques, 78
  quiet Sun, 80
  slope, 80
  stellar, 241–243
  temperature dependent factors, 74
  volume element, 76, 77
emission, and dynamo model, 247
emission, coronal, 71
emissivity, 19
empirical models, 250
  cell, 168–171
  network, 168–171
energy balance, 71, 139–141
  cool loop, 224–228
  dependence on pressure, 190
  empirical, 185–190
  global, 186–187
  in static models, 207–208
  local, 187–190

lower transition region, 229–230
stellar transition regions, 244
energy conservation equation, 200
  ionization energy, 233
  static, 185
  steady flow, 212
  with thermal diffusion, 232
energy density, from line widths, 124
energy flux
  explosive events, 147
  wave, 156
enthalpy
  flux, 139, 194
  in steady flow, 212
equilibration time, thermal, 115
excitation, 20–26
  rate equation, 21
explosive events, 141–149
  acceleration, 144
  birthrates, 145
  emission measure, 146
  kinetic energy, 147
  mass, 146
  spatial distribution, 144
extreme ultraviolet (EUV), defined, 4

Faraday's law, 197
fibrils, 64
filling factor, 176–180, 198
  role in energy balance, 186
fine structure
  in transition region, 195
  relationship to magnetic field, 68
flow models
  cool loops, 231–232
  observational consequences, 217–218
fluorescent lines, stellar, 240
flux, 19, 20
  two-level atom, 25
force free field, 199
freezing in, ion abundances, 32

Gaunt factor, 74, 75
Gauss' law, 197
geometry
  and energy balance, 192
  role in models, 195
  transition region, 76, 167, 210, 249
grains, 64
gravitational acceleration, solar, 167

heating
  and ionization, 159
  electric current, 237
  episodic, 219–224
  transition region, 187, 250
height, maximum in cool loops, 227
helioseismology, 153
hybrid stars, 240
hydrogen, neutral distribution, 66

Subject Index

hydrostatic equilibrium, 167, 199

induction equation, 198
intensity distribution, EUV, 53–54
intensity fluctuations, 60–62
intercombination lines, contribution function width, 75
ion number density, 20
ionization, 26–34
    balance, 27–31
    departures from equilibrium, 27, 112, 159, 183
    effect of diffusion, 184
    equilibrium assumption, 27
    equilibrium calculations, 29
    rate equation, 27
    time, 31, 32, 157
ionization energy, in energy equation, 233

jets, 149–150

length, cool loops, 227
limb brightening, 173–176
    observations, 58–60
    scale height, 60
limb observations, 9, 56–60
line profiles, in steady flow models, 218
loops
    cool models, 224–232
    coronal, 69
    flow models, 213–219
    scaling laws, 203–205
    transition region, 69
Lorentz force, 199
Lorentz profile, 116

macrospicules, 151
magnetic diffusion equation, 198
magnetic field
    and explosive events, 148
    and microflares, 148
    expansion, 67–68
    force free, 199
    observations, 9, 67
    size of flux elements, 196
    structure, 250
    unipolar, 68
magnetic pressure, 67
mass balance, 71
    and Doppler shifts, 137–139
mass conservation equation, 200
mass flux
    downward, 138
    in explosive events, 148
    upward, 138
mass motions, static assumption, 166
mass upflows, stellar, 240
Maxwellian distribution, departures, 166, 182–184
mean free path, electron, 182

mean molecular weight, 36
mechanical energy, in transition region, 192
microflares, 147
microturbulence, photospheric, 153
models, empirical
    average temperature-density, 11
    discrepancies with observation, 193–195
    with diffusion, 234–235
models, theoretical
    static, 203–212, 251
    steady flow, 212–219
    with turbulent conduction, 236
momentum conservation equation, 200
most probable velocity, 116
mottles, 63, 64

network, chromospheric, 52, 62
network, EUV, 51–56
    average width, 54
    contrast, 55
    Doppler shifts, 130, 133–135
    emission measure, 80
    fractional area, 55–56
nonthermal broadening, *see* velocity, nonthermal, 248–249
    stellar, 244

observational requirements, 253
Ohm's law, 197
opacity, 40–41, 166
    at low temperatures, 230
    in C IV, 119
    Lyman continuum, 180–182
oscillations
    chromospheric, 153
    photospheric, 153
    transition region, 154–156

perfect gas law, 167, 201
photoionization, 26
photosphere
    boundaries, 11
    height scale, 14
pressure
    constant assumption, 167
    scale height, 167
    stellar transition regions, 243
    transition region, 107
profile function, 115

radiation
    lower transition region, 254
    transition region loss, 139, 196
radiation loss rate, 37–40, 202
    approximate expression, 37
    at low temperatures, 230–231
    dependence on elemental abundances, 39
    table, 38
radiative decay, 18
radiative excitations, 18

radiative recombination, 18, 28
rate equations
  excitation, 21
  ionization, 27
recombination, 28
recombination time, 31, 157
reconnection, and explosive events, 148
resistivity, plasma, 197
resonance lines, contribution function width, 75
resonances, 22
Reynolds number, 198
rosettes, 63, 64

scaling laws, static, 203–205
size scales, transition region, 107
solar spectrum, 4–7
source function, from limb brightening, 175
specific heats, ratio, 200
spicules, 62–63, 65, 138, 192
  and downflows, 219
  and explosive events, 150–151
  and heating models, 223
  contribution to EUV limb emission, 174
  models for producing, 223–224
steady state assumption, 21
stellar winds, 240

temperature
  electron, 123
  excitation, 159
  ion, 115, 123
  maximum in cool loops, 226
temperature diagnostics, 41–42, 108–113
  departures from equilibrium, 160
temporal variations, in transition region, 153
thermal conduction, see conduction, thermal
thermal conductivity coefficient
  electrons, 201
  ions, 201
thermal equilibration, electron-ion, 115
three-body recombination, 18
timescales
  explosive events, 143
  ionization, 31–34
  recombinaton, 31–34
transition region
  boundaries, 13–14
  definition, 1
  height, 60
  size of structures, 55
  stellar, 238–247
  structure, 252–253
  structures, limb observations, 56
turbulence, 141
  and nonthermal broadening, 128
turbulent events, 141
two-level atom approximation, 25–26

ultraviolet (UV), defined, 4
upflows, and ionization, 158
UV emission line spectra, stellar, 239

velocity, nonthermal, 116, 117, 119
  and network, 120–121
  center to limb, 121–123
  root mean square, 116
  temperature dependence, 117–119
Voigt function, 116

waves
  acoustic, 124, 250
  acoustic flux, 125, 127
  and line widths, 124
  line profile, acoustic, 127–128
  MHD flux, 125, 127
  observational properties, acoustic, 151
  observational properties, Alfvén, 152

X-ray spectra, stellar, 243